THE CREATIVE MIND:

MYTHS & MECHANISMS

THE CREATIVE MIND

MYTHS & MECHANISMS

Margaret A. Boden

BasicBooks
A Division of HarperCollins*Publishers*

Library of Congress Cataloging-in-Publication Data

Boden, Margaret A.
The creative mind : myths & mechanisms / Margaret A. Boden.
p. cm.
Includes bibliographical references and index.
ISBN 0–465–01452–6 :
1. Creative ability. 2. Artificial intelligence. I. Title.
BF408.B55 1990
153.3′5—dc20 90—55592
CIP

First published in Great Britain
by George Weidenfeld and Nicolson Ltd.
91 Clapham High Street, London SW4 7TA
1990
First published in the U.S.A. by Basic Books

91 92 93 94 HC 9 8 7 6 5 4 3 2 1

Contents

For John Trounce
And in memory of Tony Trafford

Preface

This book offers new answers to some old questions: What is creativity? How is it possible? And could science ever explain it?

Creativity is a puzzle, a paradox, some say a mystery. Artists and scientists rarely know how their original ideas come about. They mention intuition, but cannot say how it works. Most psychologists cannot tell us much about it, either. What's more, many people assume that there will never be a scientific theory of creativity – for how could science possibly explain fundamental novelties?

Parts of the puzzle can now be put in place, for we can now say something specific about how intuition works. Creativity involves the exploration of conceptual spaces in people's minds. I describe conceptual spaces, and ways of transforming them to produce new ones, by using computational concepts. These concepts are drawn from artificial intelligence (the study of how to make computers do what real minds can do), and they enable us to do psychology in a new way.

My theme, then, is the human mind – and how it can surpass itself. We can appreciate the richness of creative thought better than ever before, thanks to this new scientific approach. If the paradox and mystery are dispelled, our sense of wonder is not.

M.A.B.
Brighton, April 1990

Acknowledgments

I am especially indebted to Gerry Martin, for his many careful comments on the entire manuscript. I am grateful to the following friends also, for their helpful comments on various sections (any mistakes are, of course, my own): Peter Bushell, Andy Clark, Ben Gibbs, Marie Jahoda, Annette Karmiloff-Smith, Stephen Medcalf, Ruth Raider, Aaron Sloman, Paul Wellings and Peter Williams.

Alison Mudd prepared the printed versions of the text, and Jacqueline Korn advised me in difficult circumstances: I thank them both. Part of the book was written during a sabbatical year granted by the University of Sussex.

Laurence Lerner kindly allowed me to reprint two of his poems from *A.R.T.H.U.R.: The Life and Opinions of a Digital Computer* (published by Harvester Press). Publisher's permission has been granted as follows: Picador, for quotes from A. Koestler, *The Act of Creation*; Duckworth, for quotes from A. Findlay, *A Hundred Years of Chemistry*; and University Press of America for quotes from H. Poincaré, *The Foundations of Science*. A few brief passages in the text are partly based on other work of mine: the sections on betrayal and the detective novelist (in Chapter 7) on my *Artificial Intelligence and Natural Man*; the discussions of BORIS (Chapter 7) and the brain-stuff argument and Chinese Room (Chapter 11) on my *Computer Models of Mind: Computational Approaches in Theoretical Psychology*; the remarks on describing noughts-and-crosses (Chapter 7) on *Minds and Mechanisms: Philosophical Psychology and Computational Models*; and the story of the compass (Chapter 11) on my paper 'Wonder and Understanding' published in *Zygon*, 1985.

As for the diagrams, I thank Harold Cohen for allowing me to reproduce the *Frontispiece* and Figures 7.2–7.9, Christopher Longuet–Higgins for Figures 5.1–5.3 (from his book *Mental Processes: Studies in Cognitive Science*), and Kyra Karmiloff for Figure 4.11. Other diagrams are reproduced with the publisher's permission as follows:

Duckworth for items from A. Findley, *A Hundred Years of Chemistry* (Figures 2.2, 4.1–4.3); Elsevier-Sequoia SA for items from A. Karmiloff-Smith's paper 'Constraints on Representational Change: Evidence from Children's Drawing', *Cognition*, 1990 (Figures 4.4–4.10); Addison-Wesley for an item from E. Charniak and D. McDermott, *An Introduction to Artificial Intelligence* (Figure 5.4); McGraw-Hill, for items from E. A. Feigenbaum and J. Feldman (eds), *Computers and Thought* (Figure 5.6); Oxford University Press, for an item from Roger Penrose, *The Emperor's New Mind* (Figure 7.1); W. H. Freeman for an item from R. C. Schank and K. M. Colby, *Computer Models of Thought and Language* (Figure 7.10); Edinburgh University Press for an item from A. Davey, *Discourse Production: A Computer Model of Some Aspects of a Speaker* (Figure 7.11); and Viking for items from D. Michie and R. Johnston, *The Creative Computer: Machine Intelligence and Human Knowledge* (Figures 8.1 and 8.2).

I am not forgetting beauty. It is because the worth of beauty is transcendent that the subtle ways of the power that achieves it are transcendently worth searching out.

John Livingston Lowes

1

The Mystery of Creativity

Shakespeare, Bach, Picasso; Newton, Darwin, Babbage; Chanel, the Saatchis, Groucho Marx, the Beatles . . . take your pick. From poets and scientists to advertisers and fashion designers, creativity abounds.

Think of a friend or relative: very likely, you can recall creativeness there, too. Perhaps no jokes up to Groucho's standards, but surely some spontaneous wit or sarcasm? Maybe they can hum their own descants to hymn-tunes, or improvise jazz on the living-room piano? And what about their ingenuity in running up a fancy-dress costume or fixing a faulty car?

Certainly, there can be disagreement about whether some idea, or person, is creative. You may draw the line at your boss's jokes or your flatmate's cooking. You may baulk at the brothers Marx or Saatchi. You may murmur that Darwin's own grandfather, among others, had the idea of evolution long before he did. You may even grumble that Shakespeare borrowed plots from Plutarch, that Bach used themes from Vivaldi, or that Picasso adapted pictures by Velásquez. But you would be hard put to deny that creativity does, sometimes, happen.

How it happens is a puzzle. This need not imply any fundamental difficulty about explaining creativity in scientific terms: scientists take puzzles in their stride.

Mysteries, however, are different. If a puzzle is an unanswered question, a mystery is a question that can barely be intelligibly asked, never mind satisfactorily answered. Mysteries are beyond the reach of science.

Creativity itself is seemingly a mystery, for there is something paradoxical about it, something which makes it difficult to see how it is even possible. How it happens is indeed puzzling, but that it happens at all is deeply mysterious.

If we take seriously the dictionary definition of creation, 'to bring into being or form out of nothing', creativity seems to be not only unintelligible but strictly impossible. No craftsman or engineer ever made an artefact from nothing. And sorcerers (or their apprentices) who conjure brooms and buckets out of thin air do so not by any intelligible means, but by occult wizardry. The 'explanation' of creativity thus reduces either to denial or to magic.

Nor does the problem concern only material creation. To define creativity psychologically, as 'the production of new ideas', hardly helps. For how can novelty possibly be explained? Either what preceded it was similar, in which case there is no real novelty. Or it was not, in which case one cannot possibly understand how the novelty could arise from it. Again, we face either denial or magic.

A psychological explanation of creativity, it seems, is in principle unachievable. It is not even clear that there can possibly be anything for it to explain. – And yet, undeniably, there is.

Philosophers and theologians noticed the paradoxical flavour of the concept of creation long ago. Two thousand years before us, they argued that creation *ex nihilo* (out of nothing) is impossible even for God. They claimed that the universe was created not only by God but also, necessarily, out of God.

This conclusion, however, does not solve the mystery. The universe apparently has ('new') properties which God does not have. So the mediaeval theologians of Christianity, Judaism, and Islam – and their successors in and after the Renaissance – painstakingly debated how it might be metaphysically possible for an immaterial God to create a material universe.

Some philosophers, today as in the past, have concluded that it is not possible at all: either there is no creator-God (and no creation), or the creator of nature somehow shares nature's properties.

But if the creator shares the creation's properties, can we really speak of *creation*? With no essential distinction between creator and created, there is nothing new, so there can be no creation. This is why Christian doctrine insists that Christ, being identical with God, was 'begotten, not created' (a phrase that occurs in a popular Christmas carol).

In short, the paradox persists.

The creation of the universe, problematic as it is, can be left to the attentions of theologians and cosmologists.

What of human creativeness, whether occasional (the boss's one witty remark) or sustained (Mozart's life-long repertoire)? Nothing could be more familiar. Psychology, surely, ought to be able to explain this?

But human creativity is problematic, too. For instance, it is not just surprising: it appears to be intrinsically unpredictable. If – as many people believe – science conveys the ability to predict, a scientific psychology of creativity is a contradiction in terms. Someone who claims that creativity can be scientifically understood must therefore show in just what sense it is unpredictable, and why this unpredictability does not anchor it firmly in the depths of mystery.

Many related problems concern just how novel a novelty has to be, to count as creative. There is novelty (and unpredictability) in randomness: so is chaos *as such* creative? There is novelty in madness too; what is the distinction between creativity and madness?

Individuals can think things which are novel with respect to their own previous thoughts. So is every banality newly recognized by an adult – and a great deal of what a young child does – to count as creative?

People can have ideas which, so far as is known, no person has ever had before. So if I remark (what no one else has ever been daft enough to say) that there are thirty-three blind purple-spotted giant hedgehogs living in the Tower of London, does that make me creative?

Suppose a chemist or mathematician has an idea that wins a coveted international award, and it later turns out that a self-educated crossing-sweeper had it first. Is this even possible, and if so does it destroy the prize-winner's creativity?

What about the recognition of novelty: if an idea is novel, why cannot everyone realize its novelty, and why is this realization sometimes long delayed? And what of social acceptance: is this relevant to creativity, and if so does it follow that psychology alone (helped by neither the sociology of knowledge nor the history of ideas) cannot explain it?

These queries have a philosophical air, for they concern not merely the 'facts' about creativity but the very concept itself. There are many intriguing factual questions about creativity – above all, just how it happens. But many recalcitrant problems arise, at least in part, because of conceptual difficulties in saying what creativity *is*, what *counts* as creative. And the factual questions cannot be answered while the conceptual paradox is raging.

One aim of this book is to arrive at a definition of creativity which tames the paradox. Once we have tamed the paradox and eliminated the mystery, creativity can sensibly be regarded as a mental capacity to be understood in psychological terms, as other mental capacities are.

This leads to my second aim: to outline the sorts of thought-processes and mental structures in which our creativity is grounded, so suggesting a solution to the puzzle of how creativity happens.

Popular beliefs about human creativity are implicitly influenced by the paradoxical nature of the concept, and are highly pessimistic about science's ability to explain it.

Indeed, 'pessimistic' is perhaps the wrong word here. For many people revel in the supposed inaccessibility of creativity to science. Two widespread views – I call them the *inspirational* and the *romantic* – assume that creativity, being humanity's crowning glory, is not to be sullied by the reductionist tentacles of scientific explanation. In its unintelligibility is its splendour.

These views are believed by many to be literally true. But they are rarely critically examined. They are not theories, so much as *myths*: imaginative constructions, whose function is to express the values, assuage the fears and endorse the practices of the community that celebrates them.

The inspirational approach sees creativity as essentially mysterious, even superhuman or divine. Plato put it like this: 'A poet is holy, and never able to compose until he has become inspired, and is beside himself and reason is no longer in him . . . for not by art does he utter these, but by power divine.'

Over twenty centuries later, the play *Amadeus* drew a similar contrast between Mozart and his contemporary, Salieri. Mozart was shown as coarse, vulgar, lazy and undisciplined in almost every aspect of his life, but apparently informed by a divine spark when composing. Salieri was the socially well behaved and conscientious expert, well equipped with 'reason' and 'art' (i.e. skill), who – for all his success as the leading court-composer (until Mozart came along) – achieved a merely human competence in his music. The London critic Bernard Levin, in his column in *The Times*, explicitly drew the conclusion that Mozart (like other great artists) was, literally, divinely inspired. If this view is correct, all hope of explaining creativity scientifically must be dismissed as absurd.

The romantic view is less extreme, claiming that creativity – while not actually divine – is at least exceptional. Creative artists (and scientists) are said to be people gifted with a specific talent which others lack: insight, or intuition.

As for how intuitive insight actually functions, romantics offer only the vaguest suggestions. They see creativity as fundamentally unanalysable, and are deeply unsympathetic to the notion that a scientific account of it might one day be achieved.

According to the romantic, intuitive talent is innate, a gift that can be squandered but cannot be acquired – or taught. This romanticism has a defeatist air, for it implies that the most we can do to encourage creativity is to identify the people with this special talent, and give them room to work. Any more active fostering of creativity is inconceivable.

But hymns to insight, or to intuition, are not enough. From the psychological point of view, 'insight' is the name not of an answer but of a question – and a very unclearly expressed question, at that.

Romanticism provides no understanding of creativity. This was recognized by Arthur Koestler, who was genuinely interested in *how* creativity happens, and whose account of creativity in terms of 'the bisociation of matrices' (the juxtaposition of formerly unrelated ideas) is also a popular view. As he put it,

> The moment of truth, the sudden emergence of a new insight, is an act of intuition. Such intuitions give the appearance of miraculous flashes, or short-circuits of reasoning. In fact they may be likened to an immersed chain, of which only the beginning and the end are visible above the surface of consciousness. The diver vanishes at one end of the chain and comes up at the other end, guided by invisible links.[1]

However, Koestler's own account of how this happens – although an advance over the pseudo-mysticism propounded by romantics and inspirationists – is no more than suggestive. He described creativity in general terms, but did not explain it in any detail.

This book takes up the question of creativity from where Koestler left it. It tries to identify some of the 'invisible links' underlying intuition, and to specify how they can be tempered and forged.

My main concern, then, is with the human mind, and how our intuition works. How is it possible for people to think new thoughts? The central theme of the book is that these matters can be better understood with the help of ideas from artificial intelligence (AI).

Artificial intelligence is the study of how to build and/or program computers to do the sorts of things which human minds can do: using English, recognizing faces, identifying objects half-hidden in shadows, advising on problems in science, law or medical diagnosis. It provides many ideas about possible psychological processes, and so has given rise to a new approach in studying the mind: 'computational' psychology.

My account of human creativity will call on many computational ideas. These can be grasped by people who know next to nothing about computers, and who care about computers even less. They can be thought of as a particular class of *psychological* ideas. As we shall see, they help us to understand not only how creativity can happen, but also what creativity is.

This claim is one which both inspirationists and romantics spurn with horror, and deride with scorn. If the source of creativity is superhuman or divine, or if it springs inexplicably from some special human genius, computers must be utterly irrelevant.

Nor is it only 'anti-scientific' inspirationists and romantics who draw this conclusion. Even people (such as Koestler) who allow that psychology might one day be able to explain creativity, usually reject the suggestion that computers or computation could have anything to do with it.

The very idea, it is often said, is intrinsically absurd: computers cannot create, because they can do only what they are programmed to do.

The first person to publish this argument was Ada, Lady Lovelace, the close friend of Charles Babbage – whose mid-nineteenth-century 'Analytical Engine' was, in essence, a design for a digital computer. Although convinced that Babbage's Analytical Engine was in principle able to 'compose elaborate and scientific pieces of music of any degree of complexity or extent', Countess Lovelace declared: 'The Analytical Engine has no pretensions whatever to *originate* anything. It can do [only] *whatever we know how to order it* to perform.'[2] Any elaborate pieces of music emanating from the Analytical Engine would therefore be credited not to the engine, but to the engineer.

If Lady Lovelace's remark means merely that *a computer can do only what its program enables it to do*, it is correct, and important. But if it is intended as an argument denying any interesting link between computers and creativity, it is too quick and too simple.

We must distinguish four different questions, which are often confused with each other. I call them Lovelace-questions, because many people would respond to them (with a dismissive 'No!') by using the argument cited above.

The first Lovelace-question is whether computational ideas can help us understand how *human* creativity is possible. The second is whether computers (now or in the future) could ever do things which at least *appear to be* creative. The third is whether a computer could ever appear to *recognize* creativity – in poems written by human poets, for instance. And the fourth is whether computers themselves could ever *really* be creative (as opposed to merely producing apparently creative performance whose originality is wholly due to the human programmer).

This book is mainly about the first Lovelace-question, which focuses on creativity in *people*. The next two Lovelace-questions are less important, except insofar as they throw light on the first. The fourth (discussed only in the final chapter) is, for the purposes of this book, the least important of them all.

My answer to the first Lovelace-question is 'Yes'. Computational ideas can help us to understand how human creativity is possible. As we shall see, this does not mean that creativity is predictable, nor even that an original idea can be explained in every detail after it has appeared. But we can draw on computational ideas in understanding in scientific terms how 'intuition' works.

The answer to the second Lovelace-question is also 'Yes', and later I shall describe some existing computer programs which, arguably, appear to be creative. (For reasons I shall discuss, programs which *unarguably* appear creative do not yet exist.)

Sometimes, these 'creations' would be worthy of admiration if produced in the usual way – whatever that is! – by a human being. One example I shall mention concerns an elegantly simple proof in geometry, which Euclid himself did not find. Nor is this appearance of computer-creativity confined to purely mathematical, or even scientific, contexts. The picture reproduced in the frontispiece sits in my office, and has been spontaneously admired by many visitors and colleagues; yet it was generated by a computer program.

The literary efforts of current programs are less impressive, as we shall see. But even these are not quite so unthinkingly 'mechanical' as (the first four stanzas of) this imaginary computer-poem, fictionally ascribed by Laurence Lerner to ARTHUR – Automatic Record Tabulator but Heuristically Unreliable Reasoner:[3]

Arthur's Anthology of English Poetry

To be or not to be, that is the question
To justify the ways of God to men
There was a time when meadow grove and stream
The dropping of the daylight in the west
Otters below and moorhens on the top
Had fallen in Lyonesse about their Lord.

There was a time when moorhens on the top
To justify the daylight in the west,
To be or not to be about their Lord
Had fallen in Lyonesse from God to men;
Otters below and meadow grove and stream
The dropping of the day, that is the question.

A time when Lyonesse and grove and stream
To be the daylight in the west on top
When meadow otters fallen about their Lord
To justify the moorhens is the question
Or not to be the dropping God to men
There was below the ways that is a time.

To be in Lyonesse, that is the question
To justify the otters, is the question
The dropping of the meadows, is the question

I do not know the answer to the question

There was a time when moorhens in the west
There was a time when daylight on the top
There was a time when God was not a question

There was a time when poets
Then I came

The third Lovelace-question arises because to appreciate originality one needs some of the abilities required for creativity itself. (In *Amadeus*, Salieri curses God bitterly for giving him enough musical talent to recognize Mozart's genius in full, but not enough to emulate it.) Indeed, being creative – as opposed to merely intriguingly crazy – requires a capacity for critical evaluation, as we shall see. Accordingly, people generally give the same answer to the third question as they do to the second.

The sceptic's resounding 'No' is expressed in another of Lerner's poems depicting 'The Life and Opinions of a Digital Computer':

Literary Criticism

Search by number	
find and gloss:	six proud walkers
	five days of Christmas
	four and twenty blackbirds
	three blind mice. Go.
three blind mice:	one noun phrase
two blind mice:	one noun phrase
one blind mice:	please repeat
one blind mice:	stop breaks rule
	align number indicators
one blinds mice:	stop breaks rule. No.
	Go obeys other rule
two blinds mice:	stop breaks rule
	align number indicators
one blinds mice:	who blinds mice?
the farmer's wife:	one noun phrase
she blinds mice:	that is aggression
see how they run:	imperative clause
see how they run:	fear-behaviour
see how they run:	using legs
the farmer's wife:	one noun phrase
cut off their tails:	more aggression
with a carving knife:	divergent thinking
did you ever see:	interrogative
did you ever see:	listener involvement
did you ever see:	listener not blind
did you ever see:	listener not mice
such a thing in your life:	listener alive
as three blind mice:	are they symbols?
four blind mice:	plight of humanity
five blind mice:	population explosion
six blind mice:	out of control
seven blind mice:	so am I
seven blind mice:	infinite loop
	see how I run
	see how I run

Lerner is a poet and critic, whose professional peers would very likely endorse his negative reply to our third question: *of course* no computer could be a literary critic, for no inhuman machine could even appreciate human creativity – still less, match it.

Computer scientists, by contrast, may give a very different answer. Alan Turing, the founder of computer science, believed that programs might one day be able to appreciate sonnets – and even to write them. Here is part of his vision of how a computer of the future might produce a literary performance indistinguishable from that of a human being (the human speaks first):

In the first line of your sonnet which reads 'Shall I compare thee to a summer's day', would not 'a spring day' do as well or better?

It wouldn't scan.

How about 'a winter's day'. That would scan all right.

Yes, but nobody wants to be compared to a winter's day.

Would you say Mr Pickwick reminded you of Christmas?

In a way.

Yet Christmas is a winter's day, and I do not think Mr Pickwick would mind the comparison.

I don't think you're serious. By a winter's day one means a typical winter's day, rather than a special one like Christmas.

This is science-fiction, indeed. However, our third question concerns not the practicality of computer-criticism, but its possibility *in principle*. Even someone who shares Turing's faith in the theoretical possibility of such a computerized sonnet-scanner may not believe that it can ever be produced in practice. (For instance, I do not.) Human readers who can scan the line 'Shall I compare thee to a summer's day?' can appreciate the rest of Shakespeare's sonnets too, and also the five poems which, with Hamlet's monologue, provided the raw material for Lerner's Arthurian anthology. A program that could do the same would need many diverse sorts of knowledge, not to mention subtle ways of realizing its importance: 'to justify the otters' simply cannot be the question.

What about computers recognizing creativity in their own performances, as opposed to ours? Could a computer have a hunch of the form 'Now I'm on the right lines!', and verify that its hunch was correct?

Well, if a computer can at least appear to be creative (as I have claimed), it must be able to evaluate its own thinking to some extent. It may not be able to recognize all its good ideas as good ones, and it may sometimes become trapped in a dead end or obsessed by a triviality. But, in that event, it would be in very good (human) company. My answer to the third Lovelace-question, then, is the same as my answer to the first two: 'Yes'.

The first three Lovelace-questions concern scientific fact and theory, and they are closely interrelated. One cannot decide whether a computer could *appear* to be creative, or to evaluate creativity, unless one has some psychological theory of what creative thinking is. So someone who is interested in the first question (the main topic of this book) will probably be interested in the other two, as well.

The fourth Lovelace-question – whether computers can *really* be creative – is very different (and, for our purposes, less interesting). It involves controversial debate about metaphysics and morals.

It raises the problem, for instance, of whether, having admitted that we were faced with computers satisfying all the scientific criteria for creative intelligence (whatever those may be), we would *in addition* choose to take a certain moral/political decision. This decision amounts to dignifying the computer: allowing it a moral and intellectual respect comparable with the respect we feel for fellow human beings.

For reasons explained in the final chapter, I would probably answer 'No' to the fourth question. Perhaps you would, too. However, this hypothetical moral decision-making about imaginary artificial creatures is irrelevant to our main purpose: understanding human creativity. For even if we answer 'No' to the fourth Lovelace-question, the affirmative answers to the first three questions can stand.

In sum: whether or not computers can really be creative, they can do apparently creative things – and, what is more to the point, considering how they do so can help us understand how creativity happens in *people*.

Programmability, of course, is not the only feature of computers which makes people doubt their relevance to creativity. For one thing, they are implemented rather than embodied. They are made of metal and silicon, not flesh and blood.

So perhaps Turing's imagined conversation is impossible after all? For it is our human experiences of sneezing and chilblains which make comparing someone to a winter's day so infelicitous, and it is our shared appreciation of good food and good cheer which make Christmas Day and Mr Pickwick such a congenial comparison.

However, this 'embodiment' objection is not directed to creativity as such, but only to certain kinds of creativity. A computer's alien indifference to winter weather may jeopardize its poetic sensitivity, but not its originality in science or mathematics.

What is more relevant, because it concerns creativity in general, is the biological fact that human intuition, *pace* inspirationists, is crucially dependent on the human brain. But computers are very different from brains.

This point is commonly remarked by those who deny that computers could even *appear* to be creative, or to appreciate human creativity. Such people claim that even to associate the concepts (never mind the experiences) of winter and Mr Pickwick requires powers of thinking or information-processing which brains have, but computers do not.

Just what these brain-supported powers of thinking are, and whether they really are utterly alien to computers and computational psychology, are questions discussed later.

We shall see that the ways in which computers are unlike brains do not make them irrelevant to understanding creativity. Even some 'traditional' work in computer science and artificial intelligence is pertinent. And the more recent 'connectionist' computers are more brain-like than the machine your gas company uses to prepare its bills, and better (for instance) at recognizing analogies – which certainly have something to do with creativity.

If the widespread human ability to recognize analogies has something to do with creativity, so do many of our other mental powers. Indeed, it is potentially misleading to refer (as I did above) to creativity as 'a capacity'. We shall see that creativity is not a single ability, or talent, any more than intelligence is.

Nor is it confined to a chosen few, for – despite the elitist claims of inspirationists and romantics alike – we all share some degree of creative power, which is grounded in our ordinary human abilities.

To be sure, creativity demands expert knowledge of one type or another – of sonnets, sonatas, sine-waves, sewing . . . And the more impressive the creativity, the more expert knowledge is typically involved. Often, the expertise involves a set of technical practices (piano-playing, designing and running experiments) which require not only years of effort but also very expensive equipment. A self-educated crossing-sweeper, no matter how intelligent, could not win the next Nobel prize for chemistry. (Perhaps a comparable prize for mathematics?)

But creativity also requires the skilled, and typically unconscious, deployment of a large number of everyday psychological abilities, such as noticing, remembering and recognizing. Each of these abilities involves subtle interpretative processes and complex mental structures.

For example, both the 'realism' of Renaissance perspective and the 'deformations' of Cubism – two truly creative movements in the development of Western painting – are grounded in the set of psychological processes that make it possible for the mind to interpret two-dimensional images as depicting three-dimensional scenes. Significantly, our knowledge of these psychological processes has been enormously advanced by computer modelling. Only a computational psychology can explain in detail how it is possible for us to see material objects as separate things, placed here or there, near or far, of this or that shape, and with surfaces sloping in this or that direction. (These cryptic remarks will be clarified later.)

How, then, can human creativity be understood? What is it, and how is it possible? Those questions are our main concern. In answering them, we shall find answers also to the four Lovelace-questions distinguished above.

Before attempting to solve the puzzle of creativity, we must first dissolve the mystery. So in Chapter 3 I offer a revised definition of creativity: one which avoids the paradoxical 'impossibility' of genuine creation, yet distinguishes mere newness from genuine novelty.

Chapter 4 discusses style and structure in the mind, and sketches some psychological factors capable of changing them. We shall see there that *exploration* is a key concept for understanding creativity, no matter what field (chemistry or musical harmony, for example) is involved.

In Chapter 5, I show why a psychological theory of creativity needs to include computational concepts. Examples of such concepts are applied to specific cases in art, science and mathematics.

Chapter 6 asks how the imagery of *The Ancient Mariner*, for instance, could have arisen in the poet's mind. It suggests that this could have resulted from mental processes similar to those which occur in brain-like, connectionist computer systems.

The following two chapters describe, and criticize, some existing computer models of creative processes. These deal with examples drawn from many different domains: pen-and-ink drawing, jazz-improvisation, story-writing, literary and scientific analogy, the diagnosis of soybean-diseases, chess, physics, chemistry, mathematics and engineering. Chapter 7 concentrates on the arts and Chapter 8 on the sciences, but the separation is a matter of convenience rather than intellectual principle: the creative processes on either side of this cultural divide are fundamentally similar. Looking at the way these programs work – and what they can and cannot do – will help us to understand how people can think creatively about such matters.

The role of chance, the relevance of unpredictability and the relation between chaos and creativity are explored in Chapter 9. We shall see that the unpredictability of (much) creativity does not put it beyond the reach of science.

The creativity of Everyman is defended in Chapter 10, whose message is that although Mozart was a super human, he was not superhuman. We are all creative to some degree – and what we can do, Mozart could do better.

Finally, in Chapter 11, I explain why we do not need the myths of inspirationism or romanticism to buttress humane values. Contrary to common belief, a science of creativity need not be dehumanizing. It does not threaten our self-respect by showing us to be mere machines, for some machines are much less 'mere' than others. It can allow that creativity is a marvel, despite denying that it is a mystery.

First of all, however, let us see (in Chapter 2) what various people – artists and scientists, psychologists and philosophers – have said about what it feels like to be creative, and about how creativity happens.

2

The Story So Far

The bath, the bed and the bus: this trio summarizes what creative people have told us about how they came by their ideas.

Archimedes leapt from his bath in joy and ran through the streets of Syracuse, crying 'Eureka!' as he went. He had solved the problem that had been worrying him for days: how to measure the volume of an irregularly shaped object, such as a golden (or not-so-golden) crown. – Friedrich von Kekulé, dozing by the fire, had a dream suggesting that the structure of the troublesome benzene molecule might be a ring. A whole new branch of science (aromatic chemistry) was founded as a result. – The mathematician Jacques Hadamard, more than once, found a long-sought solution 'at the very moment of sudden awakening'. – And Henri Poincaré, as he was boarding a bus to set out on a geological expedition, suddenly glimpsed a fundamental mathematical property of a class of functions he had recently discovered and which had preoccupied him for days.

As these (and many similar) examples show, creative ideas often come at a time when the person appears to be thinking about something else, or not really thinking at all.

Archimedes was lazing in his bath, and Poincaré was looking forward to his sightseeing trip. Kekulé was half asleep by the fire, and Hadamard was fast asleep in bed (until suddenly awakened). Marcel Proust was immersed in the most trivial of pursuits – eating a cake – when he was overcome by the recollections which led him to write his great novel. And Samuel Taylor Coleridge's poetic vision of Xanadu came to him in an opium-induced reverie. In this case, the new ideas were fleeting, and easily lost through distraction. The haunting imagery of *Kubla Khan*, with its breathtaking mixture of sweetness and savagery, would have been even richer if the 'person on business from Porlock' had not knocked on Coleridge's cottage-door.

From the creator's point of view, intuition is an enigma. Sometimes, it is experienced as a sudden flash of insight, with no immediately preceding ideas in consciousness. Hadamard is a case in point: 'On being very abruptly awakened by an external noise, a solution long searched for appeared to me at once without the slightest instant of reflection on my part.'[1]

Other times, a little more can be said. For instance, here is Kekulé's account of how, in 1865, he arrived at his insight about benzene:

I turned my chair to the fire and dozed. Again the atoms were gambolling before my eyes. This time the smaller groups kept modestly in the background. My mental eye, rendered more acute by repeated visions of this kind, could now distinguish larger structures, of manifold conformation; long rows, sometimes more closely fitted together; all twining and twisting in snakelike motion. But look! What was that? One of the snakes had seized hold of its own tail, and the form whirled mockingly before my eyes. As if by a flash of lightning I awoke.[2]

This famous fireside vision, which initiated the notion of the benzene molecule shown in Figure 2.1, will be discussed in detail in Chapter 4. Here, let us merely note that it was not an isolated incident. Kekulé had had similar experiences before (hence his reference to 'repeated visions of this kind').

Figure 2.1

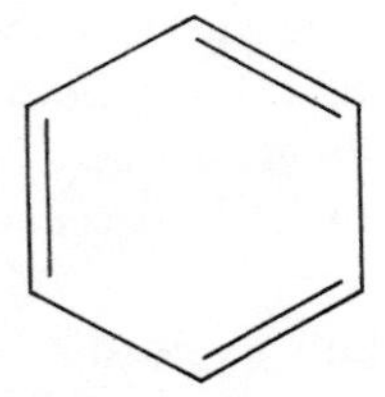

One of these had occurred almost a decade earlier. At that time, Kekulé was puzzling over how to describe the detailed internal structure of molecules:

One fine summer evening, I was returning by the last omnibus [buses again!], 'outside' as usual, through the deserted streets of the metropolis, which are at other times so full of life. I fell into a reverie, and lo! the atoms were gambolling before my eyes. Whenever, hitherto, these diminutive beings had appeared to me, they had always been in motion; but up to that time, I had never been able to discern the nature of their motion. Now, however, I saw how, frequently, two smaller atoms united to form a pair; how a larger one embraced two smaller ones; how still larger ones kept hold of three or even four of the smaller; whilst the whole kept whirling in a giddy dance. I saw how the larger ones formed a chain. . . . I spent part of the night putting on paper at least sketches of these dream forms.[3]

Partly as a result of this bus-borne reverie, Kekulé developed a new account of molecular structure, in which each individual atom could be located with respect to all the other constituent atoms. In so doing, he suggested that organic molecules were based on strings of carbon atoms. (This was independently suggested in the very same year, 1858, by the Scots chemist Alexander Couper: one of many historical examples of 'simultaneous discovery'.)

In his textbook of organic chemistry published in 1861, Kekulé represented ethyl alcohol – written today as C_2H_5OH, or CH_3CH_2OH – by the diagram shown in Figure 2.2. As you can see, the 'larger atoms' of carbon are indeed 'embracing' the 'smaller' atoms. Two hydrogen atoms are joined as a 'pair'. And the two carbon atoms are linked to form a short 'chain'.

Figure 2.2

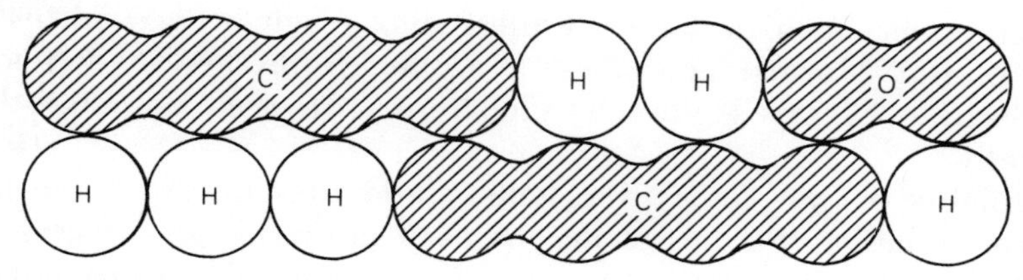

Visual imagery, obviously, was important to Kekulé (who had been a student of architecture before turning to chemistry). It was important to Coleridge, too. But many reports of creativity mention no imagery. They simply recall the sudden appearance of the solution to a problem the individual had been working on with no apparent success.

The suddenness of the solution is not its only strange feature. The answer to the prior question may be of an unexpected kind: Hadamard reports awaking with a solution 'in a quite different direction from any of those which I had previously tried to follow'. Sometimes, there appears to be no prior question – or, at least, no prior questioning. Picasso, for instance, implied that he formed no expectations, that he could advance his art without having to look where he was going: 'Je ne cherche pas, je trouve.'

Picasso spoke, as ever, in the first person. But others have disclaimed personal responsibility for the creation, or at least for significant parts of it. For instance, the novelist William Golding denies having thought of the incident in *The Lord of the Flies* when the pig's severed head speaks to the boy hiding in the bushes. He reports, instead, 'I heard it' – and remarks that at such moments 'The author becomes a spectator, appalled or delighted, but a spectator.'[4]

What are we to make of this? Although 'sudden illumination' may be a faithful description of how creativity appears to the creator, it cannot be the whole story. Intuition cannot consist merely in flashes of insight. That way, magic lies.

Magic, or perhaps theology. We no longer accept Descartes' seventeenth-century view that all human judgment is essentially unconstrained, that this intellectual freedom is the respect in which we are truly made in God's image. We should be similarly sceptical of those twentieth-century 'explanations' which appeal to an unanalysed faculty of intuition or even (as cited in Chapter 1) to divine inspiration of the creative elite. We may sympathize with Einstein's remark that Mozart was 'only a guest on this earth', or resonate to the concert-programme which declared: 'Others may reach heaven with their works. But Mozart, he comes, he comes from there!'[5] But while celebrating Mozart's glory in such terms we need not take them literally.

Insights do not come from gods – and they do not come from nowhere, either. Flashes of insight need prior thought-processes to explain them. (The aura of mystery here – if novelty is grounded in prior ideas, can it really be *novelty*? – will be dispelled by the non-paradoxical definition of creativity given in Chapter 3.)

The thought-processes in question include some conscious ones. Archimedes, Kekulé, Hadamard and Poincaré had all been thinking about their problem for many days. As for Coleridge, who had no specific 'problem' in mind when composing *Kubla Khan*, he later recalled that he had been reading this sentence just before he fell into semi-consciousness in his chair:

> In Xamdu did Cublai Can build a stately Palace, encompassing sixteene miles of plaine ground with a wall, wherein are fertile Meddowes, pleasant springs, delightful Streames, and all sorts of beasts of chase and game, and in the middest thereof a sumptuous house of pleasure, which may be removed from place to place.[6]

(Compare this sentence with these four lines from the poem: 'In Xanadu did Kubla Khan / A stately pleasure-dome decree / . . . / So twice five miles of fertile ground / With walls and towers were girdled round.') What is more, Coleridge's notebooks show that he was an exceptionally attentive reader, consciously discriminating every phrase.

Introspective reports, even those of people (like Coleridge) with a lively interest in how the creative imagination works, cannot be taken at face-value. In Chapter 10, for example, we shall see that Coleridge gave inconsistent reports of his *Kubla Khan* experience, and that some of his reports are flatly contradicted by documentary evidence. We shall see, too, that a more self-disciplined approach to introspection can show up fleeting contents of consciousness which are normally forgotten.

Even so, the many introspective reports of bath, bed and bus strongly suggest that creativity cannot be explained by conscious processes alone. Artists and scientists alike have argued that relevant mental processes must be going on unconsciously too.

Coleridge, for instance, regarded the unconscious as crucial in the creation of poetry. He was fascinated by the mind's ability to conjure up many different but surprisingly relevant ideas, and he spoke of the 'hooks and eyes' of memory. Indeed, it was because he was so interested in the unconscious associative powers of memory that he troubled to record the sentence he had been reading just before his exotic dream of Xanadu. Moreover, he saw associative memory (to be discussed in Chapter 6) as relevant not only to literary creativity but to scientific originality as well.

Poincaré, too, suggested that creativity requires the hidden combination of unconscious ideas. He distinguished four phases of creativity (which Hadamard later named *preparation*, *incubation*, *illumination* and *verification*), within which conscious and unconscious mental work figure to varying extents.

The preparatory phase involves conscious attempts to solve the problem, by using or explicitly adapting familiar methods. Often, there is no apparent success: the experience is frustrating, because seemingly unproductive.

It is during the second phase, which may last for minutes or for months, that fruitful novelties are initially generated. The conscious mind is focused elsewhere, on other problems, other projects – perhaps even on a sightseeing trip. But below the level of consciousness, Poincaré said, ideas are being continually combined with a freedom denied to waking, rational thought. (He insisted that incubation involves productive mental work, not merely a refreshing rest; some evidence in his support is given in Chapter 10.)

Next comes the flash of insight, to which – despite its unexpectedness as a conscious experience – Poincaré ascribed a significant mental history: 'sudden illumination [is] a manifest sign of long, unconscious prior work'.[7]

Finally, deliberate problem-solving takes over again, as the new conceptual insights are itemized and tested. In science and mathematics, it is natural to speak of 'verification', as Hadamard did; in the arts, the more general term 'evaluation' is better.

This fourfold analysis is no romantic hymn to the supreme majesty of the unconscious. Rather the reverse:

> A first hypothesis now presents itself: the subliminal self is in no way inferior to the conscious self; it is not purely automatic; it is capable of discernment; it has tact, delicacy; it knows how to choose, to divine. What do I say? It knows better how to divine than the conscious self, since it succeeds where that has failed. In a word, is not the subliminal self superior to the conscious self? . . . I confess that, for my part, I should hate to accept it.[8]

Far from ignoring the role of consciousness, Poincaré insisted that 'unconscious work is possible, and of a certainty it is only fruitful, if it is on the one hand preceded and on the other hand followed by a period of conscious work'.

Poincaré's account is especially well suited to mathematical and scientific creativity, where it is usual for a specific problem to be explicitly identified and explored during preparation and used as a test during verification. This is not always so where artistic creativity is concerned, for the artist may have no clear goal in mind; consider the writing of *Kubla Khan*, for instance. Even so, artists often do have a prior 'problem', or at least a project. Thus Coleridge several times announced his intention of writing a poem about an ancient mariner, and Bach composed a set of preludes and fugues systematically exploiting certain harmonic possibilities.

Moreover, poets, painters and composers commonly spend many hours evaluating their work. To be sure, they sometimes imply that no such reflection is involved: remember Picasso's 'Je ne cherche pas, je trouve.' But this does not show that Picasso used no evaluation (how did he know he had found it, when he found it?). It shows only that, in some cases, he judged the novel structure to require no modification.

'Complete' illumination of this sort is comparatively rare. Composers usually make corrections to their manuscript scores, and art-historians (using increasingly sophisticated scientific techniques) constantly discover the rejected first-thoughts of the artist, hidden under the visible layers of paint. Sometimes corrections must be minimized (if a mural has to be painted before the plaster dries), and sometimes they are impossible (when a jazz musician improvises a melody to fit a given chord sequence). Even so, the artist evaluates the production, so as to do better next time.

In short, Poincaré's four-phase theory allows that the arts and sciences achieve their innovations in broadly comparable ways.

But *what* ways, exactly? What is the unconscious work, and why must it be preceded and followed by consciousness?

Poincaré's answer was that preparatory thinking activates potentially relevant ideas in the unconscious, which are there unknowingly combined. A few are insightfully selected (because of their 'aesthetic' qualities), and then refined by conscious deliberation.

Struggling to imagine a mechanism by which these things could happen, he suggested:

> Figure the future elements of our combinations as something like the hooked atoms of Epicurus. During the complete repose of the mind, these atoms are motionless, they are, so to speak, hooked to the wall. . . .
>
> On the other hand, during a period of apparent rest and unconscious work, certain of them are detached from the wall and put in motion. They flash in every direction . . . [like] a swarm of gnats, or, if you prefer a more learned comparison, like the molecules of gas in the kinematic theory of gases. Then their mutual impacts may produce new combinations.
>
> What is the role of the preliminary conscious work? It is evidently to mobilize certain of these atoms, to unhook them from the wall and put them in swing. . . . After this shaking up imposed upon them by our will, these atoms do not return to their primitive rest. They freely continue their dance.
>
> Now, our will did not choose them at random; it pursued a perfectly determined aim. The mobilized atoms are therefore not any atoms whatsoever; they are those from which we might reasonably expect the desired solution.[9]

As for which 'new combinations' of mental atoms are most likely to be interesting, Poincaré said:

> Among chosen combinations the most fertile will often be those formed of elements drawn from domains which are far apart. . . . Most combinations so formed would be entirely sterile; but certain among them, very rare, are the most fruitful of all.[10]

Poincaré's metaphor for the mechanism of creativity is not

unpersuasive. But it sits uneasily with cases like Hadamard's, who reported finding a solution 'quite different' from any he had previously tried. If the gnat-like ideas, originally activated by conscious intent, are only 'those from which we might reasonably expect the desired solution', then how could such a thing happen?

Moreover, his image cannot explain why it is that combinations of elements drawn from different domains may be more creative, as opposed to more unnatural or absurd. Both these difficulties recall the paradox mentioned in Chapter 1, and they can be resolved only when we are clearer about the difference between creativity and mere novelty.

A third caveat is that Poincaré was not Epicurean enough. Epicurus suggested that different material atoms have different shapes, and differently shaped hooks, so that certain combinations are more likely than others. But Poincaré put the hooks on the wall, not on the atoms, and (since he ignored the shapes of atoms and hooks) implied nothing about differential affinities between ideas – still less, how normal affinities can be fruitfully overcome.

Finally, this picture of the chance impacts of undifferentiated atoms leads to a fourth problem: Poincaré's insistence on 'the automatism of the subliminal self'.

You may think it strange that I call this a 'problem', given my claim (in Chapter 1) that ideas about computing can illuminate creativity. If a computer is not an automaton, what is? Surely, someone favouring a computational account of creativity must believe in the automatism of the subliminal self (and, for that matter, of the conscious self too)?

Not necessarily, for 'automatism' can mean two different things. It may mean (positively) that the system in question functions according to scientifically intelligible principles which, together with its input-history, determine what it does. To speak of automatism in this sense is to assert that something is a principled, and perhaps largely autonomous, system (or mechanism). Alternatively, the term may be used (negatively) to deny that the system enjoys properties such as choice, discernment and judgment.

These two senses do not entail one another. They may not even apply to exactly the same class of things, because it might be possible to explain choice and judgment themselves in scientific terms. Whereas my claim was positive (that creativity can be computationally understood) Poincaré's was negative. In describing the subliminal self as 'purely automatic', he *denied* that it 'is capable of discernment; [that] it has tact, delicacy; [that] it knows how to choose, to divine'. In short, he meant that the unconscious self, like a crowd of gas-molecules, is blind.

Poincaré's view that incubation is automatic, or blindly undiscriminating, was criticized by Koestler. Although agreeing on the importance of the unconscious, Koestler did not see the random concourse of a host of separate gas-molecules as a helpful metaphor. Citing a rich and diverse harvest of historical examples, he concluded:

> The most fertile region [in the mind's inner landscape] seems to be the marshy shore, the borderline between sleep and full awakening – where the matrices of disciplined thought are already operating but have not yet sufficiently hardened to obstruct the dreamlike fluidity of imagination.[11]

By 'the matrices of disciplined thought', he meant the ordered conceptual structures which he assumed to underlie conscious reasoning.

Koestler explained creativity as the 'bisociation' of two conceptual matrices which are not normally associated, and which may even seem incompatible: 'The basic bisociative pattern of the creative synthesis [is] the sudden interlocking of two previously unrelated skills, or matrices of thought.'[12] The more unusual the bisociation, the more scope there is for truly creative ideas. Various types of unconscious thinking may be involved, including visual imagery; concrete (sometimes personal) exemplars of abstract ideas; shifting emphasis; reasoning backwards; and generating analogies of diverse kinds. In addition, he emphasized the importance of long apprenticeship and expertise, whether in science or in art.

Both Koestler and Poincaré, then, explained creativity in terms of the unconscious combination of ideas drawn from different domains. But only Koestler mentioned mental structure. He saw creativity as exploiting, and being somehow unconsciously guided by, specific conceptual matrices.

This is why Koestler rejected Poincaré's explanation as 'mechanistic'. Nothing guides one gnat to fly towards another, nothing causes one gas-molecule to move nearer to another – and nothing prevents their random dancing from falling into madness. Creativity requires more than the mere automatic mixing of ideas. Even an unguided bisociation of matrices is not enough:

> [The] rebellion against constraints which are necessary to maintain the order and discipline of conventional thought, but an impediment to the creative leap, is symptomatic both of the genius and the crank; what distinguishes them is the intuitive guidance which only the former enjoys.

As for how this intuitive guidance works, Koestler recognized the need for a detailed explanation, of which (as he was well aware) he could provide only the sketchiest outline. Discussing various scientific discoveries, for example, he said:

Some writers identify the creative act in its entirety with the unearthing of hidden analogies. . . . But where does the hidden likeness hide, and how is it found? . . . [In most truly original acts of discovery the analogy] was not 'hidden' anywhere; it was 'created' by the imagination. . . . 'Similarity' is not a thing offered on a plate [but] a relation established in the mind by a process of selective emphasis. . . . Even such a seemingly simple process as recognizing the similarities between two letters 'a' written by different hands, involves processes of abstraction and generalization in the nervous system which are largely unexplained. . . . The real achievement [in many scientific discoveries] is 'seeing an analogy where no one saw one before'.[13]

If he could not even explain how we recognize the familiar letters of the alphabet, how much more obscure is the seeing of a new analogy. What Koestler said about creativity was usually persuasive, and often illuminating. The thought-processes he described do happen, and they do seem to be involved in creativity. But because *how* they happen was not detailed, he did not fully explain how creativity is possible.

One of the strengths of Koestler's approach was that he appealed to no special creative faculty, granted only to an elite. On the contrary, he stressed the role of bisociation in everyday humour, and in the layman's appreciation (as well as the expert's creation) of innovative science, literature and art.

Poincaré, too, saw creativity as grounded in widely shared mental properties. Even Coleridge, despite his romantic stress on the guiding role of the poetic imagination, saw the origin of novel ideas as an aspect of human memory in general.

Recent psychological research (discussed in Chapter 10) supports the view that creativity requires no specific power, but is an aspect of intelligence in general – which, in turn, involves many different capacities. For example, the educational psychologist David Perkins sees creativity as grounded in universally shared psychological capacities such as perception, memory and the ability to notice interesting things and to recognize analogies.[14] (He shows, too, that more goes on at the conscious level than is usually admitted.)

What makes the difference between an outstandingly creative person and a less creative one is not any special power, but greater knowledge (in the form of practised expertise) and the motivation to acquire and use it. This motivation endures for long periods, perhaps shaping and inspiring a whole lifetime. Howard Gruber has shown (for example) how Darwin followed, and developed, a guiding idea throughout many decades.[15]

Expertise was highlighted also by the philosopher (and chemist) Michael Polanyi, for whom all skills – and all intuitive insights – are grounded in 'tacit knowledge'. Up to a point, tacit knowledge can be made explicit (in teaching children or apprentices, or in theorizing about science or art), and doing so 'immensely expands the powers of the mind, by creating a machinery of precise thought'.[16] But some unformalized knowledge always remains, and the new insights arising from it cannot be immediately captured by conscious thought. As the mathematician Carl Gauss put it: 'I have had my solutions for a long time, but I do not yet know how I am to arrive at them.'[17]

Koestler, Poincaré, Coleridge: each was a useful witness to the creative process, but none explained it in other than a vague and suggestive way.

Their accounts are highly valuable as descriptions of the phenomena to be explained. They are even useful as the beginnings of an explanation, for they indicate where to start looking in more detail at the psychological mechanisms involved. – How does bisociation of matrices actually work, and how are novel analogies recognized? What rules govern the automatism of the subliminal self, and what goes on during incubation? What are the hooks and eyes of memory, and how do they clip together?

The more recent theories, referring to everyday abilities and to expertise, raise further questions about underlying mechanisms. –How is it that people can notice things they were not even looking for? How can people recognize that two somewhat different things (two letters 'a', or two apples) fall into the same class? How is it possible for tacit knowledge to be acquired without being explicitly taught, and how can it aid creation?

All these questions must be answered, if we are to understand insight, or intuition.

People often speak as though intuition were a magical searchlight which unerringly finds its target, a special capacity for producing significant ideas. One reason is that they think of intuition as the hidden mental faculty explaining 'creativity', a concept which (as we shall see in Chapter 3) has positive evaluation built into it. Another is that creative individuals sometimes speak in this way too.

Thus Hermann von Helmholtz (a great physicist himself) said of Michael Faraday: 'It is in the highest degree astonishing to see what a large number of general theorems, the methodical deduction of which requires the highest powers of mathematical analysis, he found by a kind of intuition, with the security of instinct, without the help of a single mathematical formula.'[18] Similarly, Carl Gauss recalled, 'I had found by chance a solution, and knew that it was correct, without being able to prove it.'[19]

These reports of flashes of insight bearing the stamp of certainty on them were made with hindsight. They may be honest accounts of what someone felt at a certain moment, but we marvel at them primarily because (as we now know) the feeling marked the seed of the solution. What about the other cases?

Some new ideas so excite their creator that they take over the person's life for years to come, yet are judged by others to be worthless. Certainly, the 'others' may be wrong. Kekulé's theory of benzene was dismissed by several eminent chemists at the time as 'a tissue of fancies', and Gauguin's post-Impressionist paintings were scorned by some of his former colleagues. Even someone who tolerates a new idea may not see its implications: the President of the Linnean Society, reporting on the meetings held during 1858 – at one of which Charles Darwin and Arthur Wallace had read their papers on natural selection – said that there had been no striking discoveries that year. Indeed, this sort of thing happens so often that it requires explanation itself (it is discussed in Chapter 5). But novelty-generating cranks do exist, whose outlandish ideas are simply that – outlandish.

Obsessive ideas are sometimes so outlandish, and perhaps also disorganized, as to be judged mad. Certainly, the dividing-line between creativity and madness can be unclear. (Aristotle said that no great genius has ever been without madness, and Charles Lamb wrote to his friend Coleridge, 'Dream not, Coleridge, of having tasted all the grandeur and wildness of Fancy, till you have gone mad.'[20]) But very often we can tell the difference. Schizophrenic word-salad, for instance, is both interpretable (by such psychiatrists as Ronald Laing) and full of surprises. But it shows no psychological structure comparable to Poincaré's four phases, and it rarely produces ideas which others recognize as creative. At most, it may provide cues triggering someone else's creativity.

Other new approaches are not nurtured but discarded, and rightly so: even the most exciting idea can turn out to be a dead end.

Still others are discarded in error, their creator being blind to their significance. The mistake may be recognized later by the person concerned, or it may not. Kepler, on realizing that his long-ignored notion of elliptical orbits was not a 'cartload of dung' but 'the truth of Nature', reproached himself saying, 'Ah, what a foolish bird I have been!' Copernicus had rejected elliptical orbits too, and never realized his mistake (the relevant passage in the manuscript of *On the Revolutions of the Heavenly Spheres* was crossed out before being sent to the printers). These cases (and more) are cited by Koestler, who comments: 'The history of human thought is full of triumphant eurekas; but only rarely do we hear of the anti-climaxes, the missed opportunities, which leave no trace.'[21]

As these historical examples show, intuition (or insight) is no magical searchlight. It is not even always reliable, if what we mean by 'intuition' is *an experience* (an unexplained feeling of certainty, or significance, with respect to a newly formed idea) which people have from time to time.

It is thoroughly reliable, of course, if we take it to mean *an experience of this kind which later turned out to have been justified*. But then its 'reliability' is bogus, because we are using the word 'intuition' as we use the word 'win'. It is no more surprising that intuition (so defined) finds the right solution than that races are won by the winners. No one would ask, 'Did the winners win?', but rather 'Which runners were the winners, and how did they manage it?' The comparable question about intuition is 'Which ideas (experienced as flashes of insight) turned out to be correct, and how did the creators manage to come up with them?'

There is another question, too, comparable to 'How did the winners know that they were winning?' We need to ask, 'How can someone know that a novel idea is promising?' In other words, how are successful hunches possible?

Hunches, familiar though they are (in everyday life as in the history books), are a curious species of beast. A hunch is not merely a new idea, not merely an adventurous notion (such as might arise in brainstorming), of which one says 'Let's try this – you never know, it might work.' It is an idea which someone feels to be definitely promising, though without being able to justify that feeling rationally.

But how is it possible for an idea to strike someone as promising *before* they have checked it, given that they can say little or nothing about why they feel that way?

Poincaré attempted to answer this question by appealing to the creator's aesthetic sensibility:

> All the combinations [of ideas] would be formed in consequence of the automatism of the subliminal self, but only the interesting ones would break into the domain of consciousness. And this is still very mysterious. What is the cause that, among the thousand products of our unconscious activity, some are called to pass the threshold, while others remain below? Is it a simple chance which confers this privilege? Evidently not. . . .
>
> What happens then? Among the great numbers of combinations blindly formed by the subliminal self, almost all are without interest and without utility; but for just that reason they are also without effect on the aesthetic sensibility. Consciousness will never know them; only certain ones are harmonious, and, consequently, at once useful and beautiful. They will be capable of touching the special sensibility of the geometer [or other creative person].[22]

This answer is acceptable as far as it goes. It describes the experiences of many creative people, including Poincaré himself, and it allows for mistaken insights (Poincaré added that a false insight is one which, 'had it been true, would have gratified our natural feeling for mathematical elegance'). But it does not tell us what features make a combination seem 'harmonious', still less 'useful'. Nor does it tell us what sorts of combination (or transformation) are likely to be promising, or how their promise can be intuited.

How, for example, did Kekulé immediately recognize the tail-biting snake as potentially relevant to his problem in theoretical chemistry? How was it possible for this novel icon to trigger expectation as well as investigation? Indeed, why did this serpentine pattern arouse Kekulé's aesthetic appreciation where the others had not? Why didn't this nineteenth-century chemist see a snake in a sine-wave, or a figure-of-eight, as no less interesting? Two dream-snakes gracefully twined in a double helix might have excited Francis Crick or James Watson a century later, but there is no reason to think that Kekulé would have given them a second thought. Why not?

These questions about the origin and recognition of insightful ideas can be answered only after we are clearer about the *concept* of creativity, only after we can distinguish mere newness from genuine originality. That is the topic of the next chapter.

3

Thinking the Impossible

Alice was surprised to meet a unicorn in the land behind the looking-glass, for she thought unicorns were fabulous monsters. But having met it, she readily agreed to believe in it. We believe in creativity for much the same reason: because we encounter it in practice. In the abstract however, creativity can seem utterly impossible, even less to be expected than unicorns.

This paradox depends on the notion that genuine originality must be a form of creation *ex nihilo*. If it is, then – barring the miraculous – originality simply cannot occur.

Since it does occur, anyone loath to explain it in miraculous terms (such as divine inspiration) must find some other definition of it. If they can also show why the *ex nihilo* view is so seductive, so much the better.

Without a coherent concept of creativity, we cannot distinguish creative ideas from uncreative ones. And if we cannot do this, we cannot hope to discover the processes by which creative ideas arise. This chapter, then, tries to clarify *what counts* as a creative idea. We shall see that computational ideas can help us to understand *what creativity is* (as well as how it happens).

In the absence of magic or divine inspiration, the mind's creations must be produced by the mind's own resources. Accordingly, people who want to demystify creativity usually say that it involves some new combination of previously existing elements.

Hadamard, for example, wrote: 'It is obvious that invention or discovery, be it in mathematics or anywhere else, takes place by combining ideas.'[1] Poincaré agreed, as we have seen. Koestler conflated the explanation of how creativity happens with the definition of what it is: the bisociation of normally unrelated matrices.

In general, combination-theories identify creative ideas as those which involve unusual or surprising combinations. Some modern psychologists use the term 'statistically surprising' to define creativity, and many assume (in discussing their experiments) that the more unusual ideas are the more creative ones.

What is wrong with this? Creative ideas *are* unusual, and they *are* surprising – not least, to their originators. They may come to seem glaringly obvious ('Ah, what a foolish bird I have been!'), but they often announce themselves with a shock of surprise. Moreover, some are more surprising than others. Surely, the ideas which are 'more creative' are those which are more unusual?

One shortcoming of this definition is easily remedied. The concept of creativity is value-laden. A creative idea must be useful, illuminating or challenging in some way. But an unusual combination of ideas is often of no use or interest at all. Strictly, then, the criterion of value (to be discussed in later chapters) should be explicitly stated by combination-theories, not merely tacitly understood.

The major drawback of combination-theories, whether as definitions of creativity or explanations of it, is that they fail to capture the *fundamental* novelty that is distinctive of creative thought. It is this which makes creativity seem so mysterious, and which encourages the *ex nihilo* view in the first place. Original ideas are surprising, yes. But what is crucial is the sort of surprise – indeed, the shock – involved.

To be surprised is to find that some of one's previous expectations do not fit the case. Combination-theories claim that the relevant expectations are statistical ones. If so, our surprise on encountering an original idea must be a mere marvelling at the improbable, as when a steeplechase is won by a rank outsider.

Many examples of poetic imagery seem to fit the combination-theory account: think of Gerard Manley Hopkins' description of thrushes' nests as 'little blue heavens', or of Coleridge's description of water-snakes as shedding 'elfish' light. Scientific insights, too, often involve the unusual juxtaposition of ideas: remember the analogy Kekulé noticed between 'long rows', 'snakes' and molecules. So the combination-theory has something to be said for it (even though it does not explain how the combination comes about). But it cannot be the whole story.

To be creative, it is not enough for an idea to be unusual – not even if it is valuable, too. Nor is it enough for it to be a mere novelty, something which has never happened before. Genuinely creative ideas are surprising in a deeper way.

Where creativity is concerned, we have to do with expectations not about probabilities, but about possibilities. Our surprise at a creative idea recognizes that the world has turned out differently not just from the way we thought it *would*, but even from the way we thought it *could*.

Terms like 'unusual combination' or 'statistical surprise' do not capture this distinction. Moreover, they give no clue as to how the unusual combination could arise. (How could someone think of a thrush's nest as a little blue heaven, or of a snake as a molecule?) Creation-as-combination confuses mere abnormality, or run of the mill 'first-time' novelty, with radical originality. It does not differentiate an idea which *did not* occur before from one which, in some relevant sense, *could not* have occurred before.

Just what sense this is, needs clarification – which an explanation of creativity should provide.

On the *ex nihilo* view of creativity, the answer is easy (which is why this view can seem so attractive). The fundamental, seemingly impossible, novelty could not have occurred before the magic moment because there was nothing there to produce it. According to *The Times* columnist cited in Chapter 1, for instance, the productive source was a divine spark, without whose touch even Mozart's mind could not have achieved glory.

On a purely naturalistic view, the answer is more elusive. If *the mind's own resources* produce all its ideas, what can it possibly mean to say that an idea 'could not' have occurred before?

Certainly, the creator needs to have acquired some ideas as 'raw material', and sufficient experience to recognize and mould them. Coleridge, for example, could not have written *Kubla Khan* if he had not read the sentence quoted in Chapter 2, because its imagery is integral to the poem. Perhaps Alexander Fleming could not have discovered penicillin without finding the dirty agar-plate on the windowsill; for sure, he could not have done so without his long experience of bacteriology.

Often, however, there is no new information from the outside world, not even a chance event triggering new thoughts. Kekulé and Kepler, for instance, puzzled over their problems for a long time before finding the solutions. In such cases, where the relevant mental resources existed long before the idea appeared, surely (someone may say) it could have appeared earlier?

In what sense was Kekulé's insight about the benzene ring an event which 'could not' have happened previously? As for Kepler's idea of elliptical orbits, we have seen that it had already arisen – not only in his own mind, but in that of Copernicus too. How can one say, then, that it 'could not' have occurred earlier? Are we really talking about dates (moments in time), or something rather more subtle – and if so, what?

Before we can answer these questions directly, we must note two different senses of 'creative'. Both are common in conversations and writings about creativity, and (although the context often supports one or the other) they are sometimes confused. One sense is psychological (I call it *P-creative*, for short), the other historical (*H-creative*). Both are initially defined with respect to *ideas*, either concepts or styles of thinking. But they are then used to define corresponding senses of 'creative' (and 'creativity') which describe *people*.

The psychological sense concerns ideas (whether in science, needle-work, music, painting, literature . . .) that are fundamentally novel with respect to *the individual mind* which had the idea. If Mary Smith has an idea which she could not have had before, her idea is P-creative – no matter how many people may have had the same idea already. The historical sense applies to ideas that are fundamentally novel with respect to *the whole of human history*. Mary Smith's surprising idea is H-creative only if no one has ever had that idea before her.

Similarly, people can be credited with creativity in two senses. Someone who is P-creative has a (more or less sustained) capacity to produce P-creative ideas. An H-creative person is someone who has come up with one or more H-creative ideas.

Although H-creativity is the more glamorous notion, and is what people usually have in mind when they speak of 'real' creativity, P-creativity is the more important for our purposes.

There is an alternative definition of P-creativity, which allows for the fact that we may hesitate to credit an individual with 'creating' an idea who merely thinks of it, without realizing its significance. When Kepler first thought of elliptical orbits, he was not only surprised but dissatisfied, calling the notion 'a cartload of dung'. Only later did he recognize its importance. One might (using Hadamard's analysis of creativity) argue that he 'really' discovered elliptical orbits only when he was *inspired* by his idea and had *verified* it. Analogously, one might prefer to define a P-creative idea as a fundamental novelty (with respect to the person's previous ideas) *whose significance is recognized* by the person concerned. (Since H-creativity is defined in terms of P-creativity, there are two corresponding meanings of 'H-creativity', too.)

Each definition is defensible – and, in hotly disputed arguments about creativity, each may be passionately defended. It does not matter greatly which one we choose. Indeed, it is often unnecessary to choose between them at all, because it frequently happens that both are satisfied by the same case. What is important is that we note the psychological distinctions involved, so that we can bear them in mind when discussing the wide range of actual examples.

Kepler's idea of elliptical orbits counts as 'P-creative' on both definitions (either when he first had it or when he first valued it). But Kepler's idea counts as 'H-creative' only on the second definition (in which case, we must refuse to credit Copernicus with creativity with respect to elliptical orbits, because he never accepted them). By contrast, Kekulé's ideas about the benzene ring was, on either definition, both P-creative and H-creative.

More accurately, Kekulé's idea was H-creative *so far as we know*.

Lost manuscripts surface continually, and Sotheby's have many tales to tell of priceless paintings found in people's attics. Many creative works of art and science have doubtless been destroyed. Gregor Mendel's pioneering experiments on inheritance lay hidden for decades in an obscure botanical journal and the unread archives of an Austrian monastery. Possibly, then, someone else thought of the benzene ring before Kekulé did. (We saw in Chapter 2, after all, that Kekulé's P-novel idea about 'chains' of carbon atoms had been simultaneously P-created by Couper.)

Admittedly, nineteenth-century chemists were well aware of the rewards of establishing one's scientific priority – which today can involve unscrupulous races for Nobel prizes and international patent-rights. So it is most unlikely that any of his contemporaries – except, conceivably, a Trappist monk devoid of all self-regard and worldly ambition – had anticipated Kekulé without saying so loud and clear. (Even the Trappist monk would have needed a motivational commitment to the creative quest: if not for self-glorification, then for the glory of God. The importance of motivation in H-creativity will be discussed in Chapter 10.) It is even less likely, for reasons discussed later, that Kekulé's ideas had occurred in some previous century. Kekulé's priority is reasonably well assured.

In principle, however, ideas can be classed as H-creative only provisionally, according to the historical evidence currently available.

Because H-creativity is a historical category (many of whose instances are unknown), there can be no psychological explanation of H-creativity *as such*. Indeed, there can be no systematic explanation of it at all.

The origin and long-term survival of an idea, and the extent to which it is valued and disseminated at any given time, depend on many different things. Shared knowledge and shifting intellectual fashions are especially important (and are partly responsible for the many recorded cases of 'simultaneous discovery'). But other factors are relevant, too: loyalties and jealousies, finances and health, religion and politics, communications and information-storage, trade and technology. Even storm, fire and flood can play a part: think of the burning of Alexander the Great's library.

Iconoclasts often appeal to these social and historical contingencies in attempting to destroy, or at least to downgrade, reputations for H-creativity. One literary critic, for example, has recently argued that Shakespeare was no better a writer than several of his contemporaries, such as Thomas Middleton.[2]

He points out that Shakespeare wrote most of his plays for a single theatre-company, whereas Middleton wrote for many different theatres. It was in the Globe's interests to publish a folio of Shakespeare's plays, but there was no such commercial incentive for anyone to publish a collection of Middleton's work. Consequently, Shakespeare's plays were more widely disseminated at the time than Middleton's, and more likely to be preserved in libraries for posterity. He remarks that the Licensing Act of 1737 led to a fall in the number of new theatrical productions, so any eighteenth-century peers of the long-dead Bard would have had great difficulty getting their plays shown instead of his. He claims, too, that the 'cult' of Shakespeare over the centuries has been encouraged by people in powerful positions, in the service of various political and economic ends: recommending monarchy and preventing revolution, supporting nationalism and imperialism, encouraging tourism and promoting academic careers.

Cultural factors like these may have contributed something to Shakespeare's continuing, and widespread, reputation. There undoubtedly is a Shakespeare cult, not to say a Shakespeare industry. It does not follow that there is no significant difference between the intrinsic merits – the literary creativity – of Shakespeare's work and Middleton's. That has to be argued independently. (On this point, the relativist critic must not claim too much: if there are *no* literary values independent of cultural fashions, then one cannot say that Shakespeare was 'really no better' than Middleton, only that he was no different.)

Even someone who is convinced of Shakespeare's supreme genius must allow that many reputations for H-creativity, including his, are to some extent based on cultural factors having little to do with the intrinsic merits of the work.

Histories of science, for example, often tell a 'heroic' story – whose heroes are chosen partly for reasons over and above their actual achievement. Several different people, of different nationalities, are credited – by their compatriots – with the discovery of wireless, flying or television. Even within a single country, certain people may be described as H-creative largely because others wish to bask in their reflected glory. Detailed historical scholarship usually shows that several of their contemporaries had similar ideas, many of which even contributed to the intellectual advance.

Similar remarks could be made about many 'heroes' of the arts. In other words, the familiar adulation of 'H-creative' individuals underestimates the extent to which discovery is a *social* process.

It follows from all this that no purely psychological criterion –indeed, no single criterion – could pick out what are, by common consent, the H-creative ideas. But this does not matter: in understanding *how originality is possible*, P-creativity is our main concern.

P-creativity is crucial, too, for assessing the creativity of individual human beings, their ability to produce original ideas. An ability is a power that is more or less sustained. In other words, a person's creativity, like their intelligence, is a relatively long-lasting quality.

Granted, we often ask whether someone was being creative *at a particular point in time*, in which case we may be thinking specifically of H-creativity. Was Kepler 'really' creative when he first thought of elliptical orbits? Was Shakespeare 'really' creative in thinking of the plot of *Romeo and Juliet* (which was based on a story by Bandello), or in raiding Plutarch's *Lives* to write *Julius Caesar*? In such cases we are like the historian, having historical criteria in mind.

But creativity as a personal quality is judged (during most of the person's lifetime, if not in obituaries) primarily in terms of P-creativity. If you doubt this, consider the incident of Turing's fellowship.

As a young man in 1934, Turing (now renowned as the father of computer science, and wartime decoder of the German Enigma machine) applied for a fellowship in mathematics at King's College, Cambridge. He submitted a dissertation establishing a refinement of a well-known theorem in statistical mathematics.

The referees to whom the dissertation was sent pointed out that exactly this refinement had very recently been published by a distinguished Scandinavian mathematician. However, there was no possibility of Turing's having known of this work when he was writing the dissertation (it was not published until about the time when Turing submitted his).

Turing's paper so impressed the referees with its brilliance and originality that King's gladly offered him the fellowship. Indeed, the fact that a distinguished mathematician had also judged the result to be important was held by the existing Fellows to be a point in Turing's favour.

Were they wrong? Should they, regarding H-creativity as all-important, have refused Turing the fellowship ('Sorry, young man, but you aren't creative after all')? Or should they, grudgingly, have said: 'Well, you can have your fellowship. But you obviously aren't as creative as this other chap. It's a pity he didn't apply'? What nonsense! To offer a fellowship is not to award a prize for H-creativity. Rather, it is to bet on the Fellow's long-term capacity for producing P-creative ideas, in the hope that some of those ideas will be H-creative too. In short, the committee's decision was a sensible one.

To be sure, it is not irrelevant that the Scandinavian's result had been published only very recently. It was still surprising to most mathematicians (and Turing could not be accused of negligence for not having found it). But a gap of centuries is sometimes disregarded in judging a person's creative potential.

For instance, there is an elegant geometrical proof that the base-angles of an isosceles triangle are equal. Euclid's proof was fundamentally different, and less simple. So far as is known, the elegant proof was first discovered six centuries after Euclid by the Alexandrian mathematician Pappus – only to be forgotten in the Dark Ages, and rediscovered later. A child learning geometry at school who spontaneously came up with Pappus' proof would – justifiably – be regarded as mathematically creative. (This example was not chosen at random: we shall see in Chapter 5 that a 'Euclidean' computer program written in the 1950s produced something similar to Pappus' proof, even though its programmer did not expect it to.)

Nor is it irrelevant that only one person had anticipated Turing. Creativity is often understood in terms of the unusual, as we have seen, and H-creative ideas (which King's was betting on) are – by definition – extremely unusual. We know from experience that someone who has produced one highly unusual idea is likely to produce others. (Turing himself made fundamental contributions to computational logic, theoretical embryology and cryptology.) So the committee were justified in their decision, Scandinavians notwithstanding.

In general, however, a P-creative idea need not be unusual. It is a novelty for the person generating it, but not necessarily for anyone else. We may even be able to predict that the person concerned will have that P-creative idea in the near future, yet its being predictable does not make it any less creative. Indeed, we shall see (in the next chapter) that *every human infant* is creative, for children's minds develop not just by learning new facts but also by coming to have ideas which they simply could not have had before.

There it is again, that mysterious 'could not'. Whatever can it mean? Unless we know that, we cannot make sense of P-creativity (or H-creativity either), for we cannot distinguish radical novelties from mere 'first-time' newness.

Well, what would be an example of a novelty which clearly *could* have happened before? Consider this string of words: *priest conspiration sprug harlequin sousewife connaturality*. It probably strikes you as a random jumble of items, with no inner unity or coherent structure. And that is precisely what it is. I produced it a moment ago, by repeatedly opening my dictionary at random and jabbing my pencil on to the page with my eyes shut. But I could have produced it long since, for I learnt how to play such randomizing games as a child. Moreover, it is unintelligible nonsense: James Joyce might have done something with it, but most people could not. Random processes in general produce only first-time curiosities, not radical surprises. (This is not to deny that randomness can sometimes *contribute* to creativity – a fact discussed later.)

What about the novel suggestion made (doubtless for the first time in human history) in Chapter 1, that there are thirty-three blind purple-spotted giant hedgehogs living in the Tower of London? This, at least, is an intelligible sentence. But you could describe many more such novelties, just by substituting different English words: five compassionate, long-furred dwarf tigers sunbathing outside the Ritz . . . and so on, indefinitely.

If you were to use the terms of grammar to jot down an abstract schema describing a particular grammatical structure, you could then use it to generate an infinity of sentences – including some never heard before. For instance, the schema *determiner*, *noun*, *verb*, *preposition*, *determiner*, *noun* would cover 'The cat sat on the mat', 'A pig flew over the moon', 'An antelope eats with a spoon', and many more six-word strings (but not 'priest conspiration sprug harlequin sousewife connaturality').

A theoretical linguist would be able to provide grammatical rules describing sentences of much greater complexity, and might do so clearly enough for them to be programmed. (A program written in 1972 could parse not only 'The cat sat on the mat' but also 'How many eggs would you have been going to use in the cake if you hadn't learned your mother's recipe was wrong?') A linguist might even specify (and a computer scientist might program) a list of abstract rules capable, in principle, of generating *any* grammatical English sentence – including all those which have not yet been spoken and never will be.

The linguist Noam Chomsky remarked on this capacity of language-speakers to generate first-time novelties endlessly, and called language 'creative' accordingly. His stress on the infinite fecundity of language was correct, and highly relevant to our interests here. But the word 'creative' was ill-chosen.

Novel though the sentences about giant hedgehogs and dwarf tigers are, there is a clear sense in which each *could* have occurred before: each can be generated by the same rules that can generate other English sentences. Both you and I, as competent speakers of English, could have produced these sentences long ago – and so could a computer, provided with English vocabulary and grammatical rules.

All the 'coulds' in the previous section are *computational* 'coulds'. In other words, they concern the set of structures (in this case, English sentences) described and/or produced by one and the same set of generative rules (in this case, English grammar).

The 'and/or' is needed here because a word-string that is *describable* by the rules of grammar may or may not have been *produced* by reference to the rules of grammar. In short, computational 'coulds' come in two forms, one timeless and one temporal.

In discussing creativity – its nature and its mechanisms – we shall sometimes have to distinguish between them. One focuses on the structural possibilities defined by 'generative rules' considered as abstract *descriptions*. The other focuses on the possibilities inherent in 'generative rules' considered as computational *processes*.

To see what the difference is, consider a sequence of seven numbers $s_1, s_2 \ldots s_7$, for example the numbers 1, 4, 9, 16, 25, 36, 49. These are the squares of the first seven natural numbers (or positive integers). The sequence could be described by the rule: 's_n is the square of n' (for $n = 1$,

2 . . . 7). However, it could also be described by the rule: 's_n is the sum of the first n odd numbers' (for $n = 1, 2 \ldots 7$).

These two rules are called 'generative' by mathematicians, because they can produce, or generate, the series in question. They define timeless mappings, from an abstract schema to actual numbers. In mathematical terms, they are equivalent, since each can generate the numbers given above. (Indeed, each can generate an infinite set of numbers: *all* the squares.)

Now, instead of regarding these seven numbers as a timeless mathematical structure, consider them as a series actually written down by a friend – or by a computer. How were they produced? Perhaps your friend (or the computer) actually generated these numbers by using the first rule given above: *take the first number, and square it; add one to the first number, and square that; add two to the first number, and square that; and so on*. Or perhaps they (or it) used the second rule: *take the first number; take it again, and add the next odd number; take the result, and add the next odd number; do this repeatedly, adding successive odd numbers*. In terms of computational processes, clearly, there is all the difference in the world – especially for someone who is better (or some computer which is more efficient) at addition than multiplication. (Someone who had merely learnt this list of seven squares parrot-fashion would not have been *generating* anything, and could not go on to produce any squares they had never thought of before.)

A mathematical formula is like a grammar of English, a rhyming-schema for sonnets, or a computer program (considered as an abstract logical specification). Each of these can (timelessly) describe a certain set of structures. And each might be used, at one time or another, in producing those structures.

Sometimes, we want to know whether a particular structure could, in principle, be described by a specific schema, or set of abstract rules. – Is '49' a square number? Is 3,591,471 a prime? Is this a sonnet, and is that a sonata? Is that painting in the Impressionist style? Could that geometrical theorem be proved by Euclid's methods? Is that word-string a sentence? Is a ring a molecular structure that is describable by the chemistry of the early 1860s (after Kekulé's momentous bus-ride, but before his fireside 'dream' of 1865)? – to ask *whether an idea is creative or not* (as opposed to how it came about) is to ask this sort of question.

But whenever a particular structure is produced in practice, we can also ask what computational processes actually went on in the system concerned. – Did your friend use a method of successive squaring, or adding successive odd numbers? Did the computer use a formula capable of generating squares to infinity? Was the sonata composed by following a textbook on sonata-form? Was the theorem proved in Pappus' way or Euclid's? Did Kekulé rely on the familiar principles of chemistry to generate the idea of the benzene-ring, and if not then how did he come up with it? – To ask how an idea (creative or otherwise) *actually arose* is to ask this type of question.

We can now distinguish first-time novelty from radical originality. A merely novel idea is one which can be described and/or produced by the same set of generative rules as are other, familiar ideas. A genuinely original, or creative, idea is one which cannot.

To justify calling an idea creative, then, one must identify the generative principles with respect to which it is impossible. The more clearly this can be done, the better.

Literary critics, musicologists and historians of art examine the inherent structure of sonnets, sonatas and statues so that the nature of distinct artistic styles – and the occurrence of artistic revolutions – can be more fully appreciated. The same applies to science. Kepler's theory of elliptical orbits cannot be classed as creative without careful study of his previous thinking. Likewise, only those who know something about chemistry can appreciate Kekulé's originality.

What one knows and what one can say are, of course, two different things. We often recognize originality 'intuitively', without being able to state the previous rules and/or the way in which the new idea departs from them. Even literary critics and art-historians cannot explicitly describe every aspect of a given style. But people who sense the creativity in post-Impressionist painting, atonal music or punk rock do not have any magical insight into the aesthetics of the original. Rather, they have a general ability to recognize and compare all sorts of patterns: pandas, unicorns, apples, tables . . . (Later, we shall discuss some computational mechanisms that may underlie this ability.)

Computational concepts help us to specify generative principles clearly (many examples are given in the following chapters). And computer-modelling helps us to see what a set of generative principles *can* and *cannot* do.

Anything produced by a computer *must* (barring hardware faults) have been generated by the computational principles built/programmed into it. If a computer-model of scientific discovery (like one described in Chapter 8) produces Boyle's Law, or a 'brainlike' computer (like one described in Chapter 6) gradually learns the past tense of English verbs, irregulars and all, then their rules must have the generative potential to do so. (It does not follow that another computer – or a person – achieving comparable results need do so in the same way.)

Often, we can give a formal proof that a particular structure could or could not have been produced by a given rule-set. For instance, a computer programmed with the rules of grammar (and an English dictionary) could generate 'An antelope eats with a spoon', but not 'priest conspiration sprug harlequin sousewife connaturality'.

However, the generative potential of a program is sometimes less obvious. Lady Lovelace notwithstanding, one can be surprised by what a computer program does when it is run in a computer. By the same token, one can be surprised by the implications of a particular psychological theory. But that, too, is a topic for later chapters.

In sum, the surprise that we feel on encountering a creative idea often springs from our recognition that it simply *could not* have arisen from the generative rules (implicit or explicit) which we have in mind. With respect to the usual mental processing in the relevant domain (chemistry, poetry, music . . .), it is not just improbable, but *impossible*.

How did it arise, then, if not by magic? And how can one impossible idea be more surprising, more creative, than another? If the act of creation is not mere combination, or 'bisociation of unrelated matrices', what is it? How can creativity possibly happen?

4

Maps of the Mind

Imagine a wet Sunday afternoon, and a child clamouring for your attention. She is bored with *Snakes and Ladders*, but hasn't yet mastered chess. She has a menagerie of stuffed animals, and several dolls. She is also quite vain. (So is her brother, but he's ill in bed upstairs.) You need a game you can play together, to keep her amused for a while.

How about this? The object of the game is to make bead-necklaces, for her and the dolls and teddy-bears to wear. You give her a box of blue beads, some fine string and a bagful of ready-made necklaces (which you had the foresight to prepare) made out of red, white and blue beads.

She falls on the stuff with delight – but, you tell her, she cannot do just as she likes. The game has two rules: one about how to build new necklaces, and one about which necklaces the teddies and dolls are allowed to wear. (She, as a concession, is allowed to wear any necklace at all.)

First, you tell her how to make more necklaces. Whenever she strings a new necklace she must use a ready-made necklace (or one she has already constructed while playing the game) as a 'guide'. She may add two blue beads, no more and no less, to the guide-necklace – but she must do this in a particular way. Moreover, only necklaces of a certain kind can be used as guides.

The necklace-building rule is this: *If* she has a necklace comprising some blue beads, followed by a red bead, then some blue beads, then a white bead, and finally some blue beads *then* she is allowed to add a blue bead to both the second and the third group of blue beads. (You explain to her that 'some' means 'one or more'.)

For instance, if she picks out a ready-made necklace like this: *BBrBBBwB*, then she may string a new necklace like this: *BBrBBBBwBB*. And that necklace, in turn, will allow her to string this one: *BBrBBBBBwBBB*. (She remarks: 'Good! I'll be able to make a lo-o-o-ng necklace, even long enough for my giant panda.')

You spend a little time together practising, enlarging the collection of necklaces by using this rule. Then, just as she is about to place a newly made necklace round her teddy-bear's neck, you tell her the rule about which necklaces may be worn.

The stuffed animals and dolls are allowed to wear only necklaces whose ready-made ancestors are like this: a certain number of blue beads, followed by one red bead, one blue bead, one white bead, the same number of blue beads, and a final blue bead. In other words, the useful ready-made necklaces, for the purpose of adorning the toys, are of the form *xrBwxB* (where 'x' is some number, any number, of blue beads).

If you spend the rest of the afternoon playing this game together, you (and even she) may notice an interesting thing.

Before saying what this is, I'll give you a chance to play the game by yourself. Indeed, I strongly recommend that you do this, for reasons that will soon become clear. (You will find this preliminary playing around especially helpful if you do not think of yourself as mathematically minded.) You don't need any real beads, or string. You can use pencil and paper instead, as I did in representing the bead-sequences mentioned above.

Instead of reading the next section immediately, then, spend a few minutes playing around with the two rules. See if anything strikes you about the nature of the game (and jot your ideas down on paper).

Also, note down just what sorts of 'playing around' you actually get up to in the process. What questions do you ask yourself? And do you try to imagine any slightly different games, or are you content with this one?

Let's get back to our Sunday afternoon. You (and the child) may notice that in every doll-wearable necklace, no matter how long it is, the total number of blue beads on each side of the white bead is equal.

On seeing this, she now announces that each of her dolls and animals has its own 'lucky number'. Each one, she tells you gravely, wants its own special necklace, having its lucky number of blue beads to left and right of the white one.

You groan: you foresee tantrums, if it turns out to be impossible to make a necklace for 9, 14 or 17. You relax, however, on finding that one of the ready-made necklaces you prepared earlier is of the form: *BrBwBB*. For you realize, after a few minutes of playing around with it, that you will be able – given beads enough, and time – to make a doll-wearable necklace suitable for every doll. No matter what lucky number the child imperiously demands, you can make a necklace with exactly that number of blue beads on either side of the white. Even the green hippopotamus can have a personalized necklace.

Perhaps you disapprove of dolls, and scorn personal adornment? You favour educational games instead? Very well. You point out to the child (if she hasn't already seen for herself) that this simple game provides a way of doing *addition*.

Given any doll-wearable necklace, the number of beads in the first blue group plus the number in the second blue group equals the number in the third blue group. If, like us, she has been using pencil and paper to jot down descriptions of necklaces, she can now substitute '+' for 'r' and '=' for 'w'. When the description is of a necklace that the dolls are allowed to wear, it will now be an acceptable arithmetical equation. It may be '1 + 8 = 9', for instance, or perhaps '33 + 66 = 99'.

If you are really ambitious, and she is not too tired, you can show her that the game shares some abstract features with number theory.

For example, every doll-wearable necklace (if interpreted as an addition) represents a valid theorem of number-theory. The most useful ready-made necklace, *BrBwBB*, corresponds to '1 + 1 = 2', the simplest axiom of addition. Any integer (hence, any lucky number) can be generated by repeatedly adding *one* to any smaller integer. If she wants a necklace with nine blue beads on either side of the white one, she can be confident of doing it by working her way up to '1 + 8 = 9'.

Moreover, the number-series (the length of doll-wearable necklaces) is in principle infinite. In practice, you will have to stop: supper-time, lack of beads, no more string. But since there is no 'stop-rule', telling you to cease building whenever you have a necklace of a certain type, you could go on for ever. (She may have glimpsed this for herself, when she referred to the 'lo-o-o-ng necklace'.)

Having kept the child playing happily for some time, you are probably in a good mood. This is just as well. For the child's curiosity is by now fully engaged, and this could give you some grey hairs.

Enthusiastically, she suggests: 'Let's do "2 + 2 = 4".' Why do you anxiously search through the bag of ready-made necklaces, and just what do you hope to find?

Maybe she declares: 'When we've made the necklace for "2 + 2 = 4", I'm going to let my favourite teddy wear it.' What do you say to that?

Or she says: 'We've made a necklace saying "1 + 8 = 9", so now let's make one saying "8 + 1 = 9".' How long do you think that would take?

Perhaps she clamours: 'Let's do "1 + 1 + 1".' What then?

Or she may ask: 'If all the animals' lucky numbers were odd, and all the dolls' lucky numbers were even, what ready-made necklace would be most useful for the animals?' How would you answer?

Suppose she says, 'We've done addition-by-necklace. Now let's do subtraction-by-necklace.' Would that be a very good moment to announce '*Bedtime!*'? Or could you think up an extra necklace-building rule to meet her request?

The necklace-game is based on a formal system (the 'pq-system') defined in Douglas Hofstadter's fascinating book *Gödel, Escher, Bach.*[1] Hofstadter uses the pq-system to illustrate a host of abstract issues about the nature of generative systems, computation and representation. Our particular interest here is its ability to give a flavour of what it is like to do creative mathematics – indeed, what it is like to be creative in many different fields.

By creative mathematics, I do not mean adding 837,921 to 736,017 to get 1,573,938 (let us assume that no one has ever done that sum before). Rather, I mean producing new generative systems, new *styles* of doing mathematics.

The creative mathematician explores a given generative system, or set of rules, to see what it can and cannot do. For instance: 'Can it do addition?' 'Can it do subtraction?' 'Can it produce only odd numbers?' 'Could it have generated "365 + 1 = 366"?' 'Could it have done "5 + 7 = 12"?' 'Could it go on producing new numbers for ever?' And whenever the answer is 'No, it cannot do that,' a further query arises: 'How might the rules be changed so that it could?'

Given our discussion in Chapter 3, you will recognize all these as being *computational* 'cans' and 'coulds'. These are the foxes, when the creative mathematician is in full cry. And high-powered mathematicians are not the only ones to concern themselves with computational 'coulds'. We all do.

In the necklace story (as often in real life), it was the child who suggested doing the impossible: '1 + 1 + 1', and subtraction-by-necklace. But I expect you came up with similar questions, if you played the game yourself before reading my fictional account. (And what was Bach doing, in writing the forty-eight preludes and fugues? Why so many, and why so few?)

We all test the rules, and consider bending them; even a saint can appreciate science-fiction. We add constraints (lucky numbers?), to see what happens then. We seek the imposed constraints (only two numbers to be added) and try to overcome them by changing the rules. We follow up hunches ('Let's do subtraction, too'), and – sometimes – break out of dead-ends. Some people even make a living out of pushing the existing rules to their limits, finding all the computational 'cans' that exist: creative tax-lawyers call them loopholes (and creative tax-legislators close them).

In short, nothing is more natural than 'playing around' to gauge the potential – and the limits – of a given way of thinking. Often, this is done by comparing one way of thinking with another, mapping one on to the other in as much detail as possible. Drawing an analogy between the necklace-game and arithmetic, for instance, helps us to see what kinds of results the game can and cannot produce. (Here, we are taking analogy for granted; later, we shall consider *just how* analogies are noticed, and used.)

And nothing is more natural than trying, successfully or not, to modify the current thinking-style so as to make thoughts possible which were not possible before.

This is not a matter of abandoning all rules (there, madness lies), but of changing the existing rules to create a new conceptual space. Constraints on thinking do not merely constrain, but also make certain thoughts – certain mental structures – possible. (If you had indulgently allowed the dolls and teddy-bears to wear just any string of beads, the child could not have done addition-by-necklace.)

Sometimes, mental exploration has a specific goal: doing subtraction-by-necklace, paying less tax, finding the structure of the benzene molecule. Often, it does not.

In this, as in other ways, creativity has much in common with play. Poincaré (as we saw in Chapter 2) described the first phase of creativity – 'preparation' – as consisting of conscious attempts to solve the problem, by using or explicitly adapting familiar methods. This description fits many cases (attempts to extend the necklace-game to subtraction, for instance). But what if there is no 'problem?' Insofar as Coleridge had a consciously recognized problem in writing *Kubla Khan*, it was getting it down on paper before he forgot it – which is not the sort of problem Poincaré had in mind. Like much play, creativity is often open-ended, with no particular goal or aim.

Or rather, its goal is a very general one: *exploration* – where the terrain explored is *the mind itself*. Some explorers of planet Earth seek something specific: Eldorado, or the source of the Nile. But many simply aim to find out 'what's there': how far does that plain extend, and what happens to this river when it gets there?; is this an island?; what lies beyond that mountain-range? Likewise, the artist or scientist may explore a certain style of thinking so as to uncover its potential and identify its limits.

Explorers usually make a map, and if possible they take some ready-made map with them in the first place. Some even set out with the specific intention of map-making, as Captain Cook circumnavigated Australia in order to chart its coastline. Maps do not merely offer isolated items of information ('Here be mermaids'), but guide the traveller in various ways.

Using a map, one can return to old places by new paths: unlike Theseus, with a ball of thread to lead him out of the Labyrinth, map-bearers rarely have to retrace their steps exactly. Map-bearers can also roam throughout a circumscribed region knowing that there is something there to find: moving camp three miles to the north is rather like speaking a new sentence, or composing a new melody in a familiar musical style. The map may even indicate how explorers can get to a part of the world they have never visited. Sometimes, the map gives them bad news: to get from here to there would require them to cross an impassable mountain-range.

In short, the map is used to generate an indefinite number of very useful 'coulds' and 'cannots'. (A list of landmarks is less useful: like the parroting of the first seven square numbers, it does not generate any new notions.)

Where creativity is concerned, the maps in question are maps of the mind. These maps of the mind, which are themselves *in* the mind, are generative systems that guide thought and action into some paths but not others.

Scientific theories, for instance, define a conceptual domain which can then be explored. Find a new town, make a new dot: 'Another benzene-derivative analysed!' Follow the river, to see where it goes: 'So benzene is a ring! What about the other molecules found in living creatures?' Identify the limits: 'Is the genetic code (whereby DNA produces proteins) the same in all living things?'

Theoretical maps help scientists to seek, and find, things never glimpsed before (like the source of the Nile). For example, Mendeleyev's 'periodic table' suggested to nineteenth-century chemists that unknown elements must exist, corresponding to specific gaps in the table. Theoretical maps also help those who want to know 'how to get there from here'. Thus general knowledge of chemical structure suggests specific pathways for synthesizing substances (including some never encountered before). And where several maps already exist, scientists may chart the extent to which they correspond. The periodic table was originally based on the observable properties of elements, but it was later found to map on to a classification based on atomic number.

A new theoretical map may not be universally welcomed, because yet-unseen spaces can be hard to imagine. An idea essentially similar to the periodic table had been proposed three years before Mendeleyev championed it, and one prominent scientist had sarcastically enquired whether the proposer had thought of classifying the elements according to their initial letters.

This historical incident reminds me of a *New Yorker* cartoon showing Einstein running his hands through his hair in frustration, his blackboard covered with crossed-out symbols. He was muttering to himself, '$e = ma^2$ – No! . . . $e = mb^2$ – No, that's not quite right!' He was exploring all right, but – as the cartoonist was reminding us – some pathways, some structures, are less likely to be fruitful than others.

Artists explore their territory in similar ways, mapping and re-mapping as they go. Impressionist painters asked what could be done – and sought the limits of what could be done – by representing three-dimensional scenes in terms of patches of colour corresponding to the light reflected from the scene. Claude Monet's series of paintings of the Japanese bridge and the water-lily pond at Giverny, for example, became stylistically ever more extreme.

When the rules have been well and truly tested, so that the generative potential of the style is reasonably clear, boredom and/or curiosity invite a change in the rules. Pointillistes such as Georges Seurat and Paul Signac turned from patches of colour to mere points (and also explored the possibilities of a strictly limited palette). Painters (such as Paul Gauguin) who had initially been trained as Impressionists threw over that style for another one –and so on, and on.

In Western music we find an analogous exploration: a continual definition, testing and expansion of the possibilities inherent in the scale. As we shall see, the progression from Renaissance music to the wilder reaches of Schoenberg is intelligible as a journey through this musical space.

Sometimes, structural aspects of a mental map are consciously accessible. Chemists explicitly seek to analyse theoretically related compounds. Bach's forty-eight preludes and fugues were a systematic exploration (and definition) of the possibilities and internal structure of the well-tempered scale. The pointillistes deliberately decided which colours they would use. And Charles Dickens knowingly exploited English grammar in describing Ebenezer Scrooge as 'a squeezing, wrenching, grasping, scraping, clutching, covetous old sinner'.

(Strictly, this literary conceit does not fit the strong definition of creativity given in Chapter 3. Dickens was exploring grammar, but not transforming it. However, although sevenfold strings of adjectives *could* have occurred before, many readers had not realized the possibility. Dickens showed that there are more things in grammatical space than were normally dreamed of in their philosophy.)

Often, however, the map is not (or not fully) accessible to consciousness. There is no special mystery about this. Most of our abilities depend, wholly or in part, on mental processes hidden to conscious awareness. Nor are psychologists the only people who attempt to discover mental maps by indirect, non-introspective means. The theoretical linguist, the musicologist, the literary critic and the historian of science or of art: all these seek to chart the different styles of thinking employed (consciously or not) in their chosen domain.

In many ways, then, mental exploration is like the land-based variety. But there is one crucial difference. Mental geography is changeable, whereas terrestrial geography is not.

Admittedly, both are affected by chance events and long-term alteration: serendipity and volcanoes, senility and continental drift. But only the mind can change itself. And only the mind changes itself in selective, intelligible ways. The 'journey through musical space' whose travellers included Bach, Brahms, Debussy and Schoenberg was a journey which not only explored the relevant space but created it, too. And this creation, like all creation, was selectively constrained. (Indeed, controversial though it was while it was occurring, with hindsight it seems virtually inevitable.)

In short, only the mind can change the impossible into the possible, transforming computational 'cannots' into computational 'cans'.

Consider Kekulé, for instance, dozing at the fireside. The chemical theory he started from was the current orthodoxy of 1865: that all organic molecules are based on strings of carbon atoms. (He himself, as noted in Chapter 2, had originated this theory some eight years earlier.)

The organic chemist's job was to discover, by experiment, the nature and proportions of elements in a given compound, and then to describe a carbon-string providing for just those elements, in just those proportions. The description must be internally coherent (judged by the rules governing atomic combination within molecules), and must also fit the compound's behaviour observed in the test-tube.

This job had been carried out successfully for ethyl alcohol, and for many ('aliphatic') organic compounds like it. With respect to benzene, however, it had not. Indeed, the contradictions involved suggested that no such description could possibly be found.

The difficulty concerned the valency of carbon. Chemists had known since 1852 that atoms have strictly limited powers of combination, or valencies. In developing his string-theory in 1858, Kekulé had taken carbon to have a valency of four, and hydrogen a valency of one. (This is why the carbon atom in Kekulé's diagram of Figure 2.1 is represented as being four times larger than the hydrogen atoms.)

A carbon atom in a string of carbon atoms uses up one unit of its valency by being connected to another carbon atom. So it has three units left over for combination with non-carbon atoms if it is at the end of the string, and two if it is inside the string (accordingly, ethyl alcohol is often represented as CH_3CH_2OH rather than C_2H_5OH).

Experimental evidence had shown that the benzene molecule was made up of six carbon atoms and six hydrogen atoms. However, a sixfold string of carbon atoms should have a total of fourteen hydrogen atoms, not six.

The problem could not be solved by saying that some of the carbon atoms within a benzene molecule are linked by double or triple bonds, because this was inconsistent with the compound's chemical properties. (If benzene contained redundant bonds, these should be able to 'capture' monovalent atoms like chlorine or fluorine; but the chlorine atoms remained safely uncaptured.)

There seemed to be no way in which benzene could be given a chemically intelligible molecular structure. Having wrestled with this problem for many months, Kekulé then had the experience described in Chapter 2:

I turned my chair to the fire and dozed. Again the atoms were gambolling before my eyes. This time the smaller groups kept modestly in the background. My mental eye, rendered more acute by repeated visions of this kind, could now distinguish larger structures, of manifold conformation; long rows, sometimes more closely fitted together; all twining and twisting in snakelike motion. But look! What was that? One of the snakes had seized hold of its own tail, and the form whirled mockingly before my eyes. As if by a flash of lightning I awoke.[2]

It is not clear just what was going on here, even at the conscious level. This is not (though it is sometimes reported as) a description of seeing pictures in the fire, of interpreting flames as snakes. But was it a dream, or a reverie? Did Kekulé see snakes, or merely shapes that reminded him of snakes – or both? Did he see a snake biting its own tail, or merely a snaky shape closing on itself – or, again, both?

As we shall see, it does not matter greatly which account we choose (and his experience could just as well have been one of seeing pictures in the fire). Whichever phenomenological description of the fleeting, shifting images is correct, the *closure* of one of the 'snakes' was clearly the significant feature.

But why? Would an image of a circular hoop, a familiar children's toy in Kekulé's time, have been as fruitful? Would Kekulé's insight have happened even sooner if he had visualized sine-waves, which are distinctly serpentine mathematical forms? A snake seizing its own tail is admittedly surprising – but so what? Why was it so arresting that Kekulé suddenly awoke?

As for the analogous ring-structure, contemporary chemistry deemed it *impossible*. How, then, could Kekulé – aided by a tail-biting snake – generate the idea of a ring-molecule? His remark that 'This time the smaller groups kept modestly in the background' suggests that he was specifically concerned with the carbon atoms, leaving the hydrogen atoms to take care of themselves. But, faithful to the chemical theories of the time, his vision initially depicted 'long rows' of atoms. How could this string-vision give way to an image like a tail-biting snake?

There are various ways in which the seminal snake-image could have arisen in Kekulé's mind. Much as a map suggests a number of pathways by which to reach a given place, so a highly complex computational system – such as a human mind – allows for many differing routes by which to generate a certain structure, or idea.

For example, let us suppose two things: that the notion of an *open curve* already existed in Kekulé's mind, and that the ability to *consider the negative* of various concepts was also available. Each of these suggestions is independently plausible, as the next two sections will show.

As for 'open curve', this is a topological notion. Topology is a branch of geometry that deals with neighbour-relations; it includes the theory of knots. A topologist describing an egg will tell us, for instance, that an ant crawling on the surface would have to cross the 'equator' in getting from one end to the other. Shape and size are irrelevant: if you squeeze a plasticine egg, its topological properties do not change; and a reef-knot is a reef-knot, however thick the thread with which it is tied, and however loose the knot may be.

An *open curve* has at least one end-point (with a neighbour on only one side), whereas a closed curve does not. An ant crawling along an open curve can never visit the same point twice, but on a closed curve it will eventually return to its starting-point. These 'curves' need not be curvy in shape. A circle, a triangle and a hexagon are all closed curves; a straight line, an arc and a sine-wave are all open curves.

We can infer from Kekulé's own report of his reverie that 'row' and 'molecule' were actively associated within his mind. Very likely, given his chemical knowledge, 'string' was activated also. *If* the notion of an open curve already existed in Kekulé's mind, *then* this concept could have been activated too – perhaps via 'string', and maybe also 'knots'. (We are taking mental association for granted here, postponing questions about how it works to later chapters.)

This topological classification might indeed have been present in his mind. Kekulé could not have explicitly encountered topology in his studies of mathematics, since it had not yet been developed (by Poincaré). However, there is some evidence (due to the research of Jean Piaget) that every young child's thought and action is implicitly informed by basic topological notions such as this. If this suggestion is correct, then Kekulé possessed the potential – in the form of sufficient computational resources – for classifying a string-molecule as an open curve.

However, a tail-biting snake is not an open curve but a closed one. So where did this idea come from? There are several possibilities.

One is that Kekulé's habit of visualizing groups and rows of atoms relied on a general ability to transform two-dimensional shapes, which simply *happened* to come up with a closed curve. This, by the attraction of opposites which is common in mental association, could then have activated 'open curve', and so on to 'molecule'.

A second possibility is that his visualizing (when thinking about theoretical chemistry) was normally limited by a 'strings-only' constraint, which – in exploratory mode – he subconsciously dropped, with the result that closed curves could now arise.

And a third possibility is that he used the heuristic of 'considering the negative'.

Like Molière's character who unknowingly spoke in prose, we all use heuristics, whether we have ever heard the word or not. (Assuredly, then, Kekulé used them too.) In other words, heuristics form part of the computational resources of our minds.

A heuristic is a form of productive laziness. In other words, it is a way of thinking about a problem which follows the paths most likely to lead to the goal, leaving less promising avenues unexplored. Many heuristics take the current map of conceptual space for granted, directing the thinker on to this path rather than that one. Others change the map, superficially or otherwise, so that new paths are opened up which were not available before.

The study of heuristics as an aid to creativity has a long history. Pappus of Alexandria, our fourth-century friend encountered in Chapter 3, mentioned them in his commentary on Euclid. The twentieth-century mathematician George Polya has identified a wide range of heuristics, some so general that they can be applied to many sorts of problem.[3] Advertising agents and management consultants use them continually, and often explicitly, in trying to encourage creative ideas by 'brainstorming', or 'lateral thinking'. And several educational programmes, used in schools around the world, use heuristics to encourage exploratory problem-solving.[4]

Most heuristics are pragmatic rules of thumb, not surefire methods of proof. Although there is a reasonable chance that they will help you solve your problem, they can sometimes prevent you from doing so. For example, 'Protect your queen' is a very wise policy in chess, but it will stop you from sacrificing your queen on the few occasions where this would be a winning move.

Some heuristics are domain-specific, being the 'tricks of the trade' used by the skilled expert. These may be of no interest if one is concerned with a problem of a different type. 'Protect your queen', for instance, is useless as advice on how to play poker.

Others are very general, being useful in exploring fields ranging from drama to dressmaking. For instance, consider how Polya's heuristics might help a casting-director or a couturier.

Polya recommended (among other things) that one break the unsolved problem into smaller problems that are easier to tackle, or try to think of a similar problem which one already knows how to solve. He suggested that, if you are stuck, you should ask: What is the unknown? What are the data? Have I used all the data? Can I draw a diagram? Can I draw up a plan for solving the problem step by step? Can I restate the problem? Can I check the result? Can I work backwards? Can I modify a familiar solution-method to make it suitable for this case? All these heuristics apply to problems outside mathematics – even to casting a play or designing a dress.

One very general heuristic is *consider the negative*. In other words, one way of getting a new slant on a problem is to negate some aspect of it, whether consciously or unconsciously. When this heuristic is applied to a structural aspect of the problem, as opposed to a mere superficial detail, it can change the conceptual geography in one step.

(Negating a constraint is not the same as dropping it: wanting *any non-red sweet* is not the same as wanting *any sweet, whatever its colour*. However, dropping constraints is a general heuristic, too. We have already seen that Kekulé's visualization of this problem could have been transformed by dropping the 'strings-only' constraint. Likewise, dropping Euclid's sixth axiom – that parallel lines meet at infinity – led to a fundamentally different, non-Euclidean geometry.)

If Kekulé, like most of us, had this widely used negation-heuristic potentially available in his mind, it is plausible that it would be applied to the *spatial* aspects of the problem. For Kekulé's conscious task was to identify the spatial disposition of the atoms, and his semi-conscious reverie was specifically focused on variations of spatial form. Using this heuristic, he could have passed directly from 'open curve' to 'closed curve', thus establishing a connection between the concepts of molecules and closed curves.

Alternatively, the negation-heuristic might have been applied to the 'strings-only' constraint on visualization, generating a positive propensity (not just a computational possibility) for producing images of closed curves.

Kekulé's introspective account suggests that the 'snake' which 'seized hold of its own tail' was actually conceptualized during his experience (as opposed to being merely his way of describing the imagery when he wrote about it later). This is quite possible, because the initial visual imagery could have reminded him of snakes ('reminding' is a common source of creativity). In that case, he may well have 'seen' a snake biting its tail.

Our previous discussion shows that he did not need to think of snakes, as such, in order to come up with the idea of ring-molecules. However, for reasons explained below, *this specific interpretation* of his image could have helped to generate his new chemical insight.

Other scenarios – different computational processes in Kekulé's mind – are possible, too. For instance, Kekulé's unconscious might have associated the chemical term 'string' with the everyday sense of the word. The whole point of household string is that it can be knotted – or, in other words, closed. So a further association, of string with closure, is entirely plausible. ('String' might have arisen from 'chain', a concept already connected with similar visions in Kekulé's mind, and which also is associated with attachment and closure.) The notion of a closed string, given that Kekulé's visual imagery of 'long rows' was active at the time, could have triggered the idea of a snake biting its tail.

Again, Kekulé's insight could have been grounded in serendipity (a concept discussed in Chapter 9). That is, the snake-image might have arisen through mental processes unconnected with his problem, then to be seized on by Kekulé as an interesting idea.

For instance, perhaps (as suggested above) one of the 'long rows', unselectively twining and twisting, simply happened to twist into a snakelike form. Maybe Kekulé, taking a country walk before his fireside nap, had encountered a snake (dead or alive) with its tail in its mouth. Or his intuition might have been triggered by his dozing dream-memory of a snake in a painting by Hieronymus Bosch. Conceivably, the snake-image might have arisen from physiological causes, thanks to some foodstuff or hallucinatory drug which upset Kekulé's mental processes. The possibilities are endless.

Any of these things *could* have happened. Associations much more complex than this are commonplace, and only to be expected in a rich conceptual network such as the human mind. Poetry thrives on them. Even Freud's most far-fetched dream-interpretations describe associative processes that could have happened.

We shall never know just how Kekulé's idea arose in his mind. Indeed (for reasons discussed in Chapter 9), the origins of a novel idea can rarely, if ever, be known in detail. But for our purposes this does not matter. To show *how it is possible for creative ideas to occur at all*, it is enough to show that specific ideas *could* have arisen in specifiable ways. Similarly, it does not matter which phenomenological description of Kekulé's experience is the most accurate, provided that we understand, in principle, how each of them might trigger and/or mediate the creative transformation of string into ring.

No matter how it arose, Kekulé's snake-idea could have appeared *significant* only to someone with the relevant knowledge ('fortune favours the prepared mind'). Like all of us, Kekulé was able to recognize analogies. Analogy is crucial to much creative thinking, in science and the arts, and later we shall ask what sorts of computational mechanisms might underlie it. Here, we are interested in why Kekulé thought the analogy between snakes and molecules to be an exciting one.

His hunch, his feeling that this new idea was promising, was based on his chemical expertise. A tail-biting snake is surprising not only because it is rarely seen, but also because it is *an open curve that unexpectedly becomes a closed one*. It is the latter feature which proved so arresting, which awakened Kekulé 'as if by a flash of lightning'.

A snake that bites its tail thereby effects a topological change – the same change that is involved in passing from string-molecule to ring-molecule. And topology, by definition, is concerned with neighbour-relations.

As a competent chemist, Kekulé knew that neighbour-relations are important. The relations between the constituent atoms determine a substance's chemical behaviour. Moreover, valency – the concept causing difficulty in this specific case – concerns what and/or how many neighbours a given atom can have. Indeed, the fact that the snake's tail does not merely *touch* its mouth, but *fits into it*, may have reflected Kekulé's concern with valency: which atoms 'fit' with which.

A topological change in molecular structure *must* alter the neighbour-relations. And the greater the change in neighbour-relations, the more different the valency-constraints on the molecule may be. The change from an open curve to a closed one is more fundamental than merely shifting the positions of individual atoms or atom-groups within a string. So the experimental results which could not be explained by string-molecules *might* be consistent with ring-molecules.

It is thus intelligible – indeed, eminently reasonable – that Kekulé should have felt so excited by his strange idea, even before he had had the opportunity to verify it.

A hoop-image might not have captured his attention. A hoop is a hoop is a hoop – no topological surprises there. Indeed, in a pamphlet of 1861 Joseph Loschmidt had speculated that benzene has a core consisting of the six carbon atoms, arranged in two layers of three, and had drawn a diagram representing this two-layer core as a circle. If Kekulé had seen Loschmidt's diagram, it might well have been reactivated during his own thinking about benzene. But if so, it made no conscious impression on him; it triggered no startled 'But look! What was that?' Moreover, a perfect circle would not encourage someone to focus on the individual carbon-carbon links, because (being free of twists and turns) it cannot distinguish them.

As for snaky sine-waves, these are open strings, just as 'long rows' are. Except as associative stepping-stones to snakes, sine-waves would not have been much help.

The verification itself was not straightforward, for (as you may have noticed already) merely changing from string to ring does not solve the original problem. In short, this was a case where an initial hunch required further modification before its promise could be fulfilled.

Simply closing up a carbon–carbon string, and assuming one hydrogen atom for each carbon, would mean that each carbon atom would use up only three of its four valency units. In other words, six carbon atoms now seemed to require twelve hydrogen atoms. Kekulé therefore suggested an additional rule-change, namely that three of the six carbon–carbon links involved two valency-units, not one: see the hexagonal ring in Figure 4.1. This, in turn, raised the question of which links were bivalent and which monovalent: all carbon atoms being equivalent, how could the molecule 'decide'? Kekulé replied, in effect, that it could not. That is, he suggested that the single and double bonds oscillate spontaneously, so that a given molecule switches back and forth ('all twining and twisting in snakelike motion', perhaps?) between the two forms shown in Figure 4.2.

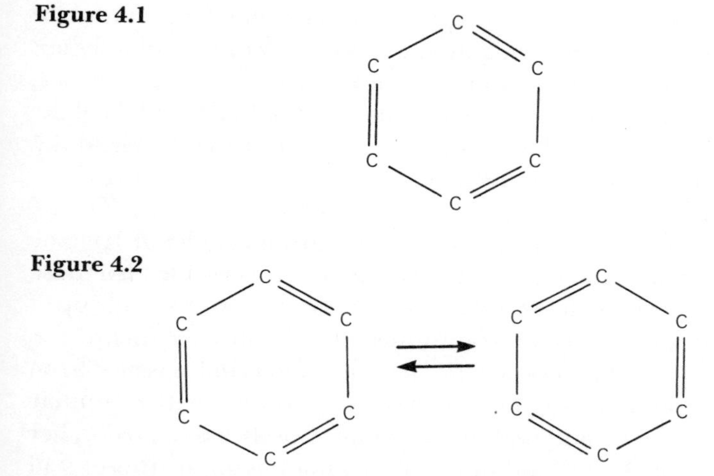

Figure 4.1

Figure 4.2

Because of these complications, Kekulé's theory was not immediately accepted by all chemists. Several other ways of dealing with the 'extra' valency-bonds were suggested, as shown in Figure 4.3 (notice that the first two of these accept Kekulé's 'flat' hexagon, whereas the third substitutes a three-dimensional prism). After much argument and experimentation, Kekulé's oscillation-hypothesis was vindicated.

There is a final twist, however: it is now accepted, on the basis of wave mechanics and electron-beam analysis, that the second diagram in Figure 4.3 – which represents the intermediate stage between Kekulé's two forms shown in Figure 4.2 – is the best *general* representation of the benzene molecule.

Figure 4.3

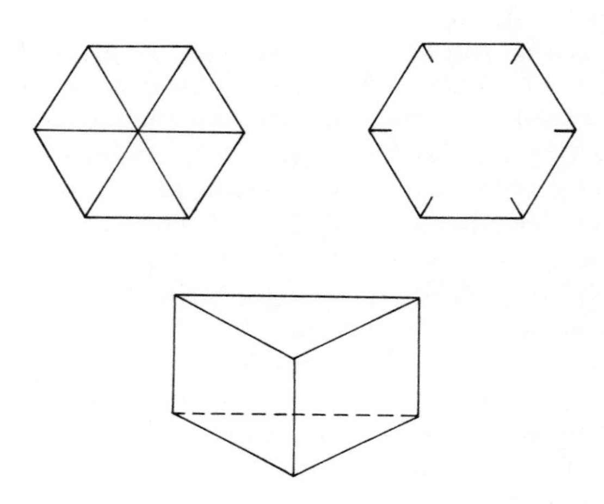

(The self-educated crossing-sweeper mentioned in Chapter 1 might have seen a tail-biting snake lying on the road, but would probably not have known enough chemistry to make the connection. Even if he did, he might not have noticed – still less solved – the residual problem of the missing valency-units. And lacking access to a laboratory, he could not have done any follow-up experiments.)

By this exploratory step, passing from 'string' to 'ring', Kekulé created the possibility of a whole new science: aromatic chemistry (the study of the benzene derivatives). Indeed, he made possible also those areas of chemistry which deal with molecules based on rings made up of different numbers of atoms, and/or atoms of different elements.

The child playing the necklace-game, who suggested passing from addition to subtraction, was taking the first step on a rather similar road, except that the second notion was *already* closely associated in her mind with the first. If (instead of cravenly announcing bedtime) you tried to accommodate her by thinking up a new rule, one which would make necklaces shorter instead of longer, you were taking further steps along that road. You may well have had some P-creative ideas (that is, some psychologically creative ideas). But this route had already been well explored by others, so H-creativity (historical novelty) was not in question.

Kekulé's idea, by contrast, created a historically new conceptual space. Its features (and potential inhabitants) were similar to, yet intriguingly different from, those he had encountered before.

Artists, too, create new conceptual spaces. By dropping the tonal constraints that had informed all Western music since the Renaissance, composers at the dawn of the twentieth century passed from one conceptual space to another – which had not existed previously.

Or rather, it had existed only implicitly, as a potential within tonality itself. Given the exploratory impulses and rule-changing heuristics available to musicians (as to the rest of us), the limits of tonal space would inevitably be respected, tested, modified and finally abandoned. As Charles Rosen puts it in his study of Schoenberg, tonality contained the seeds of its own destruction.[5]

In tonal music, every piece – be it sonata, symphony or chorale – must be structured around a key triad, the tonic chord. This defines the 'home' key, from which the piece starts and to which (perhaps after being diverted to one or more other keys) it must return. The listener whose mind contains a map, or tacit representation, of the same conceptual space feels satisfied when the piece ends on the tonic chord of the home key.

Within this musical convention, a theme-melody introduced in the home key may be transposed into the foreign key, only to reappear on the return to the home key. Often, fragments – or even whole octaves – of the (home or foreign) scale feature in the composition, not as mere melodic decoration but as a reminder (and a reinforcement) of the central role of the key in mapping the musical journey concerned. The space of all possible tonal music is structured, certain keys being regarded as more or less akin to, or distant from, each other.

Initially, certain routes to home were preferred, or even required. One would not normally consider flying direct from New York to Cambridge (England), but would fly via London instead. Similarly, in the early stages of tonality the composer would not step directly on to the final tonic chord from *any* other chord, but (for example) from the dominant of the key in question. However, much as a wealthy businessmen, bored with waiting at airports, might buy a private plane to fly directly to Cambridge, so a composer, bored with the current constraints, might try using some other chord as the tonic's predecessor.

The development of tonal music over several centuries involved an exploration (and a continuing extension and redefinition) of the harmonic steps by which a melody could progress from one movement, phrase, chord or note to the next. It also explored the major diversions, the modulations into one (eventually, more than one) 'foreign' key before coming home. Again, there was a structured space of chordal successions by which to modulate from one key to another. To arrive at a far-distant key (as defined within this space), one would normally pass through successive neighbouring keys.

By the late nineteenth century, the pathways between these modulations had become progressively shorter, the jumps from one key to another more and more daring. An early Mozart sonata might have only one modulation, into a near-neighbour key sustained throughout the second movement. By the time of Brahms and Chopin, the modulations were happening between phrases, or even within them.

Accordingly, the notions of 'current key' and of 'approved chordal succession' became increasingly problematic. In time, the musician's exploratory impulse demanded that *every* combination of notes should be allowable on the way to the conclusion of the piece.

Even so, the conclusion was still (in the first decade of the twentieth century) thought of as a resting-place, a harmonic consonance defining journey's end. In dropping even this constraint, Schoenberg stepped right out of the (by now much deformed) conceptual space of tonality, into a new field governed by different rules, in which ideas of consonance and modulation could not even be expressed.

(Schoenberg continued to test and transform the structure of musical space, developing further generative rules during his career. For example, he sometimes ruled that any given piece should use the full range of the chromatic scale. At other times, he avoided – or insisted upon – the repetition of certain notes, or series of notes.)

Rosen remarks that Schoenberg, despite being recognized as the 'destroyer' of tonality, took tonality more seriously than did any of his avant-garde contemporaries.

That is, he saw that the conventions about modulations, approved cadences, repetitive themes and concluding consonance were not arbitrary aspects of tonality. Given the fundamental tonal concept of a home key, they were intelligible, mutually coherent constraints. Being defined at less fundamental levels than the home key, they could be more easily transformed or even dropped. Nevertheless, they were all part of one coherent generative system.

Some avant-garde composers had dared to cast aside all of these constraints except one: that the piece should end with a consonance. But the authority of the home key had been so undermined, and its identification made so unclear, by the main body of the piece (which was virtually chaotic, in tonal terms) that the final 'consonance' was merely a superficial concession to convention – like a fig-leaf. Musical honesty, no less than musical daring, demanded that the fig-leaf be discarded.

In sum, it was always possible that creative exploration (broadly analogous to that outlined in relation to the necklace-game) would transform, and eventually break out from, the space of musical possibilities defined by tonality. Indeed, people being the intrepid explorers that they are, it was – as Rosen says – inevitable. It might not have been Schoenberg, and it might not have occurred in 1908. But it had to happen, some time.

It could not, however, have happened in the sixteenth century. Schoenberg had to have predecessors. One cannot utterly reject tonality (or the string-theory of molecules) without knowing what it is.

There must have been previous composers, determined to push modulation to its limits (although these need not have been Brahms, Chopin, Debussy and Scriabin). Before them, there must have been trail-blazers mapping the structural skeleton of tonal music (as Bach was doing in his 'Forty-Eight'). And before them, someone must have taken the first exploratory steps in using what is recognizably a scale (or proto-scale), even if it is not yet clearly defined in contrast with the preceding style (the 'modes' of mediaeval music).

The conceptual space defined by tonality, its potential considered as a generative system, was so rich that mapping it would inevitably take a long time. In fact, it took several centuries.

If, by some miracle, a composer had written atonal music in the sixteenth century, it would not have been recognized as creative. To be appreciated as creative, a work of art or a scientific theory has to be understood in a specific relation to what preceded it. We saw in Chapter 3 that a creative idea is one which surprises us because it could not have happened before. This is a computational 'could', to be interpreted in relation to a particular way of thinking, or generative system.

Only someone who understood tonality could realize just what Schoenberg was doing in rejecting it, and why. Similarly, only someone who knew about string-molecules could appreciate Kekulé's insight. To recognize a structural novelty, one needs a structured mind. So Salieri, a highly competent and much admired musician, might indeed have cursed God for giving him talent great enough to appreciate Mozart's genius better than anyone else.

This explains why the ignorant (that is, the inexperienced) fail even to recognize originality, never mind welcome it. But the knowledgeable often spurn it, too. Like many creative artists, Schoenberg was not universally appreciated even by his peers, many of whom reviled his music as a cacophony.

To some extent, such reluctance to accept new artistic ideas springs from a temperamental and/or socially comfortable unadventurousness. But it is due also to the difficulty (at least for adult minds) of making truly fundamental conceptual shifts.

An H-creative idea sometimes involves such a radical change in mental geography, requiring such a different sort of map to represent the new range of computational possibilities, that many people's minds cannot immediately accommodate it. And artists, of course, cannot bludgeon their critics with independently verifiable facts. They can only seek to persuade them that the mental exploration is intelligible, and therefore – like the climbing of Everest – justified for its own sake.

Scientific creativity is probably less often rejected, at least after the initial period, than artistic originality. (Although Kekulé had his critics too, as we have seen, 'a tissue of fancies' being the chemist's equivalent of 'cacophony'.) This is due partly to the general public's realization that scientific judgment requires specialist knowledge, as against their belief that the arts should be immediately intelligible. In addition, the scientist's (fourth-phase) verification involves experimental methods specifically designed to achieve universal agreement. And the paths connecting today's conceptual territory to yesterday's usually remain well trodden: old experimental results are not abandoned, but explicitly integrated within current scientific theory.

Exploration is involved even in non-revolutionary scientific research, which Thomas Kuhn called 'puzzle-solving'.[6] This term should not be interpreted too dismissively. Not all everyday science is like a (non-cryptic) crossword-puzzle, where the rules are clear and unchangeable. Like the child puzzling over the necklace-game, scientists starting from given rules try to test and bend them, to find their potential and their limits and, often, to extend their scope. 'Normal' science makes not only factual additions and corrections (as a cartographer sadly deletes *Here be mermaids*), but also theoretical modifications – from 'strings' to 'rings', for example. In short, it is creative, for in exploring its home-territory it changes the maps it inherits.

What Kuhn called 'revolutionary' science is more daring. It draws new charts of such a different kind that the traveller may seem to have lost all bearings.

On the rare occasions when the conceptual change is so radical that it challenges the interpretation of all previous experiments, scientists can differ as bitterly as any art-lovers. They have to judge alternative explanations not by a single test but by many different, and partially conflicting, criteria – some of which are not even consciously recognized. Rational argument alone may not solve the dispute, which is partly about what should count as scientific rationality. Kuhn even remarks that 'revolutionary science' succeeds because the (still unpersuaded) old scientists die.

The young, and/or those outside the professional institutions, are better able to conceive of a new conceptual space. Albert Einstein, for instance, was a young man working in a patent office when he wrote his revolutionary paper on relativity. For a variety of psychological reasons, the young – whether in science or in art – tend to be less inhibited about changing the generative rules currently informing their minds.

The very young are even better. They have an unjaded curiosity, generating all manner of mental adventures challenging the limits of the possible: lucky numbers, '1 + 1 + 1', subtraction-by-necklace. – And the very, very young are best of all.

A young child's ability to construct new conceptual spaces is seldom appreciated even by its doting parents. All human infants spontaneously transform their own conceptual space in fundamental ways, so that they come to be able to think thoughts of a kind which they could not have thought before. Their creative powers gradually increase, as they develop the ability to vary their behaviour in more and more flexible ways, and even to reflect on what they are doing.

Creativity, whether in children or adults, involves exploration and evaluation. The new idea must be compared to some pre-existing mental structure, and judged to be 'interesting' by the relevant criteria. A person who can evaluate their own novel ideas will accept them or (sometimes) correct them, but will often be unable to explain in just what way they are interesting. We need to understand why this is, and how the ability to map and explore aspects of one's own mind develops in the first place.

The psychologist Annette Karmiloff-Smith claims that when children (and adults) practise new skills, they spontaneously develop explicit mental representations of knowledge they *already* possess in an implicit form.[7] These representations arise on several successive levels, each time enabling the person to exploit the prior knowledge in ways that were not possible before. The person progresses from a skill that is fluent but 'automatic' (being varied only with much effort and limited success) to one that can be altered in many ways.

To test her theory, and to study the constraints on flexibility in the mind, Karmiloff-Smith designed experiments on skills such as drawing, speaking and understanding space or weight. Let us take drawing as our example.

The experiments involved over fifty children aged between four and eleven. Each child was asked to draw 'a house'. Then the first drawing was removed, and the child was asked for 'a house that does not exist' (or 'a funny house', 'a pretend house', 'a house you invent', and so on). Similarly, the children were asked to draw a man and then a funny man, and an animal followed by a pretend animal. In every case, the way in which an individual child went about drawing each picture was carefully observed.

Figures 4.4 to 4.9 show a few of the drawings that were done. You can see that non-existent houses, pretend animals or funny men differ from their real counterparts in a number of ways. In other words, they can be regarded as 'interesting' for different reasons.

There are cases where there is a change in the shape, or the size, of component elements; so a door is spiky, and a head is tiny or square. There are cases where the shape of the whole thing is changed; so we have houses like tripods or ice-cream cones. Sometimes, elements are deleted, giving doorless houses or one-legged men. Other times, extra elements are inserted, resulting in many-headed monsters of various kinds. (Notice that almost all of the extra elements in Figure 4.7 are truly *inserted* into the structure of the drawing, not added after the thing has been drawn as a normal whole.) Again, there may be changes in the position or orientation of elements, and/or of the whole thing; we see doors opening into mid-air, an arm and a leg switched, and a house upside-down. Finally, there are cases where the extra elements come from a different category of thing; so a man is given an animal's body, and a house is given wings.

These imaginative changes do not happen at random. The flexibility of the drawing-skill – the creative range – depends on the age of the artist. All the children could draw (real) houses, men and animals fluently: their sketches were done quickly and effortlessly. But drawing funny houses, or men that do not exist, required them to alter their usual drawing-method. The younger children found this difficult,

Figure 4.4

Shape and/or size of elements changed (ages are in years, months).

Figure 4.5

Shape of whole changed (ages are in years, months).

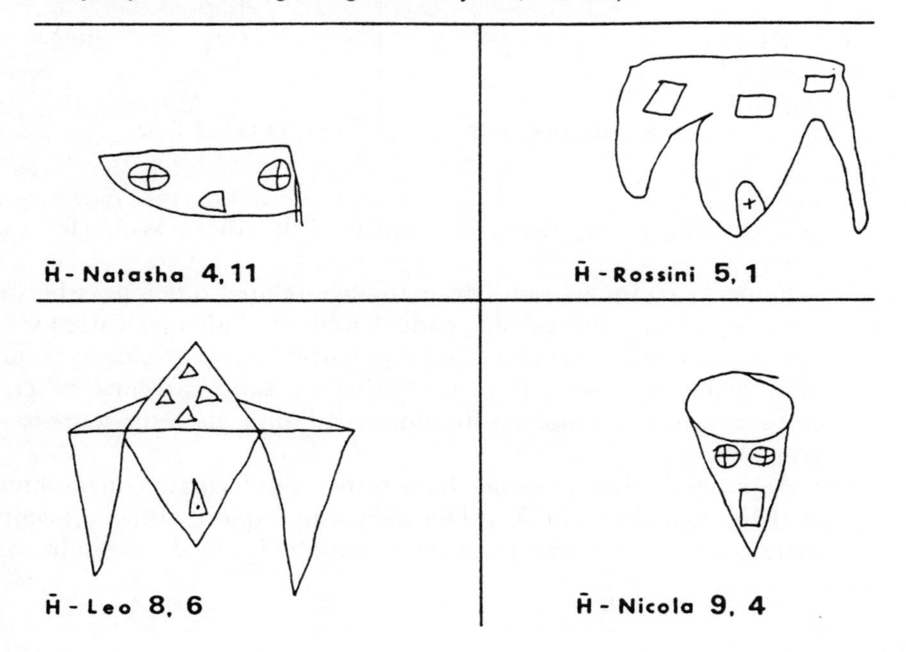

Figure 4.6

Deletion of elements (ages are in years, months).

M̄-PETER 5, 3

H̄-MARY 5, 3

M̄-ANNA 5, 6

M̄-BETTIAH 10, 2

H̄-JOSHUA 8, 8

M̄-VALERIE 9, 0

being unable to vary their drawings in all the ways possible for a ten-year-old.

Figure 4.10 shows some dramatic age-related differences between the types of transformations made. Children of all ages varied size of shape, and deleted elements. But the eight- to ten-year-olds were much more likely to insert elements (whether same-category or cross-category), or to change position or orientation, than the four- to six-year-olds.

Apparently, four-year-olds have rather uninformative mental maps of their own drawing-skills, for they can explore these conceptual spaces only in very superficial ways. A path here or there can be made

Figure 4.7

Insertion of new elements (ages are in years, months).

Figure 4.8

Position/orientation changed (ages are in years, months).

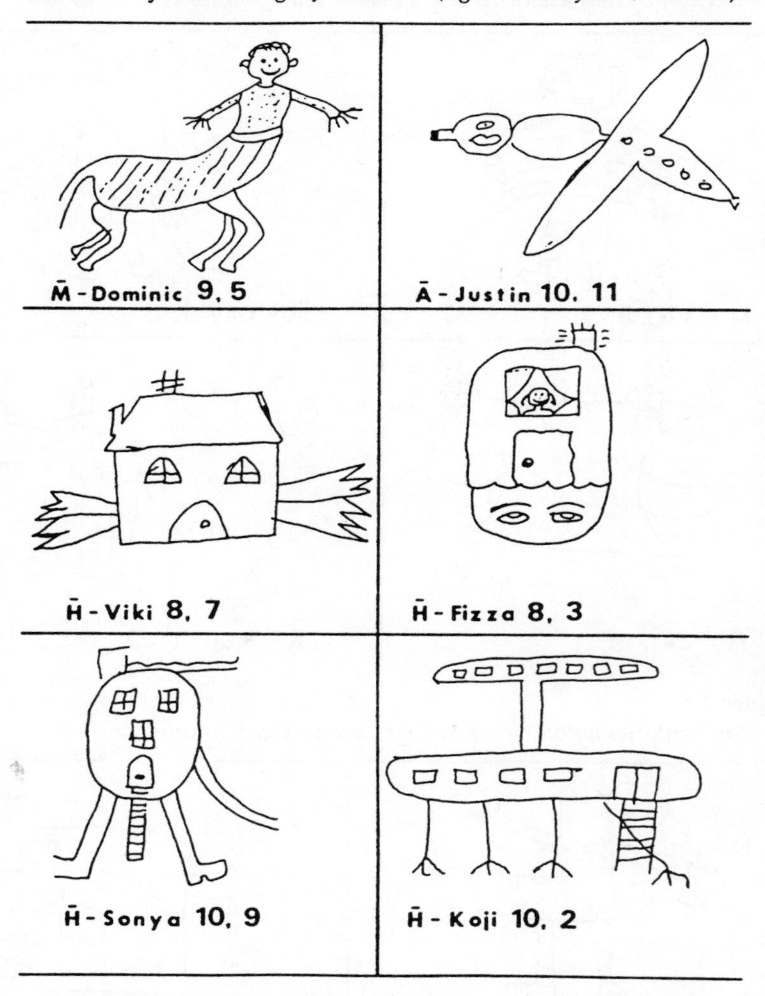

Figure 4.9

Insertion of cross-category elements (ages are in years, months).

wider or more crooked, or sometimes (under special conditions, described below) deleted altogether. But *this* path cannot be inserted into *that* one. Orientation and position are fixed: it is as though a river flowing from north to south could be made wider, and more meandering,but could not be made to run from east to west – nor transferred from the Himalayas to the Alps. And there are no pathways made up of alternating stretches of river and road: such a mixture appears to be inconceivable.

The ten-year-olds, by contrast, can explore their mental territory in all these ways. Their mental maps seem to make the necessary sorts of distinction, for they can ask what would happen if the river became part-road – and they can make it happen.

Figure 4.10

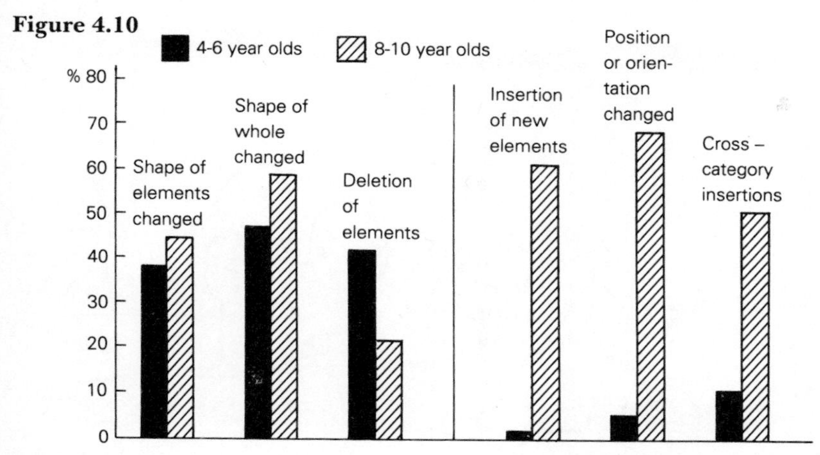

These changes in imaginative power come about, according to Karmiloff-Smith's theory, because children develop explicit representations of knowledge they already possess implicitly. In other words, the skill is *redescribed* at a higher level. (The earlier representation is not destroyed, and is still available for routine use if required.) Whereas implicit knowledge can be used but not explored, explicit descriptions allow an activity to be transformed in specific ways.

The four-year-olds are constrained by a fixed, 'automatic' sequence of bodily actions, which they can vary only in very limited ways. They can draw a man with ease (quickly, and without hesitation or mistake), but not a two-headed man. Because their knowledge of their own drawing-skill is almost entirely implicit, they can generate hardly any variations of it. They are like someone who knows how to reach a place by following a familiar route, but cannot vary the route because they have no map showing how the various parts relate to one another. The oldest children represent their skill in a much more explicit manner, and as a result can produce drawings which the younger children cannot.

To say that very young children 'cannot' draw a variety of funny men may seem strange. As Figure 4.11 shows, a three-year-old may draw a man in several ways in the same session, omitting different parts each time. But the reason is that three-year-olds have not yet mastered the relevant skill: they try to draw real men, but their drawings often turn out to show non-existent men. Precisely because the child has not yet developed any automatic procedure allowing men to be drawn with fluency, a variety of quasi-men are drawn. This is novelty, but not creativity: the variety here is due to incompetence, not to controlled flexibility. Karmiloff-Smith is interested in the latter, and in the

Figure 4.11

representational changes that occur *after* the child has achieved mastery. That is why no three-year-olds were included in her experiment.

At the first level of redescription, a drawing-skill (already mastered as a rigid sequence of bodily actions) is represented in the mind as a strictly ordered sequence of parts – for instance, head-drawing or limb-drawing parts. This sequence must be run from beginning to end, although it can occasionally be stopped short. Variable properties of the parts (like size and shape) are explicitly marked, allowing for certain sorts of imaginative distortion: heads can be made square, or arms very small (See Figure 4.4). But the relation between the parts is represented only implicitly, depending on their order in the drawing-procedure as a whole.

Consequently, body-parts are dropped by four-year-olds only rarely, and then only if they are at the *end* of the procedure. So an arm or a leg may be dropped (by a child who normally draws it last of all), but the head – because it is usually drawn first – is almost never left out.

This first-level description of the basic bodily skill can generate funny men, but their inner structure does not vary. It does not allow for repetition of a part 'at the same place' within the sequence. Nor can it generate any reordering of the parts, or the insertion of a part into the sequence. To be sure, five-year-olds were able (if asked) to draw 'a house with wings'. But they did this by *adding* the wings to the completed house, not by interrupting their drawing of the wall-lines so as to *insert* the wings smoothly into the picture. Similarly, of the very few who mixed categories without being asked to do so, all added the foreign parts after the main item had been drawn normally.

In short, very young children *cannot* insert extra elements into their drawings. When Karmiloff-Smith asked five-year-olds to draw 'a man with two heads', she found (as she predicted) that most could not do so. Typically, they would draw two heads and then attach a body-with-arms-and-legs to each head. If they were dissatisfied with the result, they would start again – but they succeeded only after very slow and elaborate efforts. They even found it difficult to *copy* drawings of two-headed men. They seemed to have an inflexible man-drawing procedure, which had to be run straight through. The first line of the head triggered the rest of the procedure, and it was impossible to go back and correct what had been done.

At the next level of description, the structure of the skill is mapped as a list of distinct parts which can be individually repeated and rearranged in various ways. The ordering-constraint is relaxed (though not dropped): a single part can now be deleted from the middle of the process, without disrupting the rest of the drawing. As a result, we see much more flexible behaviour. Funny men with extra arms are drawn, and houses are spontaneously given wings. But the flexibility is limited: there are no two-headed men, for instance, and the wings are still added to the house rather than inserted in it.

As the representation develops further, many of the structural relations between the (second-level) parts come to be explicitly mapped, and can then be flexibly manipulated. Subroutines – even some drawn from different categories – can be perfectly inserted into a drawing-procedure, the relevant adjustments (such as interruption of lines) being made without fuss. For the first time, we see funny men with two heads and three legs, fluently drawn with no rubbings-out. Also, we see parts of one representation being integrated into another, so that man and animal, for instance, are smoothly combined.

Evidently, ten-year-olds can explore their own man-drawing skill in a number of systematic ways. They can create funny men by using general strategies such as distorting, repeating, omitting or mixing parts chosen from one or more categories. In effect, their conceptual space has more dimensions than the conceptual space of the four-year-olds, so they can generate a wider – and more interesting – range of creations.

Conscious self-reflection, Karmiloff-Smith suggests, is a result of many-levelled representations of this kind. Her evidence is largely drawn from experiments on language, in which she studied not only the development of children's speaking-skills but also their growing ability to comment on what they were doing.[8]

She found, for example, that when children first use the words *the* and *a* correctly, they have no insight into what they are doing. If asked to pick up 'a watch' from a group of objects, they will (correctly) pick any of the several watches, whereas if asked for 'the red watch' or 'my watch', they pick a particular one; and they use the right word in describing what they picked up. But they cannot reflect on their behaviour, to say what the relevant difference is. They cannot even do this when they start correcting themselves for wrong uses of the word ('the watch' when there are two watches, for example). At first, they can deal with the mistake only by providing the correct form of words, not by giving the general principle involved.

Later, however, they can explain, for instance, that if there had been two watches on the table it would have been wrong to say 'I picked up the watch' (or 'Pick up the watch'), because the listener would not know which one was meant.

In short, when children first use words correctly, they do not know what they are doing. Even when they become able to correct their own mistakes, their self-monitoring does not imply a conscious grasp of the structure of the conceptual space concerned. Self-reflective insight into their own speech, enabling them to explain why they use one word rather than another, comes later. It is made possible by successive redescriptions of their pre-existing linguistic skill, comparable to those we have discussed with respect to drawing.

Very likely, these sorts of spontaneous redescription go on in adult minds too, constructing conceptual spaces on many different levels. As the successive representations multiply, the skill becomes more complex and subtle – and more open to insightful control.

Psychological evidence on how adults learn to play the piano, for example, is reminiscent of Karmiloff-Smith's data on children's skills. At first, players make many mistakes, different each time. Then they are able to play the piece through perfectly, but only by starting at the beginning. Only much later can they produce variations on a theme, or smoothly insert passages from one piece into another. Similarly, newly literate adults find it much easier to delete the final phoneme of a word than the first one. (Illiterate adults cannot selectively drop phonemes at all.)[9]

With the help of language (including technical languages like musical notation), many domain-specific structures can become accessible to consciousness. Even so, people's ability to reflect on their own skills is limited. (Bach knew full well what he was doing when he defined the home keys for the 'Forty-Eight', but he could not have said so clearly how he invented the 'funny' fugues.) Many mental processes are mapped at some higher level in a person's mind, enabling them to explore their space of skills in imaginative ways. But since not all aspects of skill are represented at a consciously accessible level, creative people usually cannot tell us how their novel ideas came about.

Adults and children alike, then, spontaneously construct maps of the mental processes going on in their own minds. These maps are used in exploring the many-levelled conceptual spaces concerned, which can even involve a transformation of the territory itself.

So far, so good. But this intuitive talk of 'maps' and 'conceptual spaces' is very vague. Anyone hoping for a scientific explanation of creativity must be able to discuss mental spaces, and their exploration, more precisely.

For example, how can one describe a conceptual space, as opposed to merely listing its inhabitants? Indeed, how can one know that a particular idea really is situated in one conceptual space, rather than another? How can one say anything about such a space before it has been fully explored? And how can one compare conceptual spaces, in terms of differences that are more or less fundamental?

In terrestrial exploration, we know the broad outlines of what to expect. We know how to tell whether a river flows into this sea or that one, and we understand the difference between fundamental geological change and mere surface-erosion. But conceptual spaces seem decidedly more elusive.

To make matters worse, they can be created afresh at any time, so that radically new maps (not just an extra dot on the old one) are required. How can one capture such ephemera? And how can one clearly distinguish various ways of transforming one conceptual space into another? In short, how can there possibly be a scientific explanation of creativity?

Fortunately, a science already exists in which conceptual spaces can be precisely described: namely, artificial intelligence (AI). Artificial intelligence draws on computer science (and also on psychology, linguistics and philosophy) in studying intelligent systems *in general.* The new computational concepts developed by AI-workers are now influencing more traditional disciplines – not least psychology.

For example, the work on children's drawing described above was strongly influenced by such ideas. Karmiloff-Smith compares the four-year-olds' drawing-skills (and early skills in general) to the kind of computer program called a 'compiled procedure': a sequence of instructions whose internal structure is not accessible to any higher-level routine, so cannot be altered. Again, she sums up her theory of representational change by saying that 'knowledge embedded in procedures gradually becomes available, after redescription, as part of the system's data-structures'. In other words (using the terms introduced in Chapter 3), a generative system which actually produces structures comes to be treated as a generative system which describes them. The 'passive' system can be examined, and transformed, by an 'active' system on a higher level.

Artificial intelligence can not only describe conceptual spaces: it can actively explore them, too. A program, when run on a computer, is a *dynamic* chart of the computational space concerned, a rambler's map that actually goes on a ramble. And the footsteps by which this exploration takes place can be precisely identified, as specific computational processes.

This is not to say that all the necessary sorts of footwear have yet been designed. To ramble through the human mind we may well need seven-league boots. But current AI can take us further than a Japanese lady's sandals, as we shall see.

5

Concepts of Computation

There are maps, and there is map-making. A map can be inadequate in many different ways: a village omitted, a river misplaced, the contours too coarse-grained to help the rambler. But inadequate maps do not show map-making to be a waste of time, and clumsy contours do not prevent 'contours' from being a useful concept. Furthermore, maps improve, as cartographers increase their geographical knowledge and think up new ways of charting it. By today's standards, a mediaeval *Mappa Mundi* is inadequate on many counts. But a map it is, even without Mercator's projection or lines of latitude. In short, if we want a systematic description of our landscape, map-making is the relevant activity.

Likewise, there are computer programs, and there are computational concepts. The programs discussed in this book have many weaknesses. But if current programs fail to match human thought, it does not follow that the theoretical concepts involved are psychologically irrelevant. Indeed, many of these concepts are more precisely defined versions of psychological notions that existed years before AI came on the scene.

Moreover, AI-workers are as creative as anyone else. New computational concepts, and new sorts of program, are continually being developed. For instance, 'neural network' systems (described in Chapter 6) illuminate certain aspects of creative thought better than early AI-work did. This is a new science, barely half-a-century old. But even a computational *Mappa Mundi* is a worthwhile achievement, as we shall see.

Two computational concepts with which to map the changing contours of the mind have already been introduced (in Chapters 3 and 4 respectively): *generative system* and *heuristic*.

Just as English grammar allows for sentences even about purple-spotted hedgehogs, so other generative systems implicitly define a structured space of computational possibilities. Heuristics are ways of selectively – insightfully – moving through this space and/or of transforming it, sometimes by changing *other* heuristics. 'Protect your queen' directs you into some chess-paths and away from others. And 'consider the negative', if applied at a relatively deep level of the generative system, can transform the space so fundamentally that very different sorts of location are created and many previous locations, indeed whole regions, simply cease to exist.

For instance, Kekulé's consideration of closed (not-open) curves led to the discovery of a wide range of molecular structures. And Schoenberg, on considering the exploratory potential of the chromatic – *not* the tonal – scale, lost the possibility of writing music in a home-key. (Or, rather, he provisionally surrendered it. Human minds are able to contain several maps for broadly comparable territory – and, gentle Reader, some creative art achieves a mischievous or ironic effect by juxtaposing signposts drawn from very different styles.)

Neither of these concepts is new. Generative systems and heuristics had been studied by mathematicians (Pappus and Polya, for instance) long before they were used in AI, and heuristics were investigated also by Gestalt psychologists such as Carl Duncker and Max Wertheimer. Indeed, the first heuristic programs were partly based on the insights of Polya, Duncker and Wertheimer.

However, AI can provide dynamic processes as well as abstract descriptions. Consequently, it can help us to compare generative systems clearly, and to test the computational power of individual heuristics in specific problem-solving contexts.

The 'dynamic processes', of course, are functioning computer programs. A program (together with an appropriate machine) is what computer scientists call an *effective procedure*. An effective procedure need not be 'effective' in the sense of succeeding in the task for which it is used: doing addition, recognizing a harmony, writing a sonnet. All computer programs are effective procedures, whether they succeed in that (task-related) sense or not.

An effective procedure is a series of information-processing steps which, because each step is unambiguously defined, is guaranteed to produce a particular result. (It can include a randomizing element, if a particular step instructs the machine to pick one of a list of random numbers; but the following step must specify what is to be done next, depending on which number happened to be picked.) Given the appropriate hardware, the program tells the machine what to do and the machine can be relied upon to do it. As Lady Lovelace would have put it, the computer will do precisely what the program orders it to do.

The early AI-workers who first defined heuristics as effective procedures developed the related concept of a *search-space* (one example of what in Chapter 4 was called a conceptual space). This is the set of states through which a problem-solver could conceivably pass in seeking the solution to a problem. In other words, it is the set of conceptual locations that could conceivably be visited.

'Conceivably' here means 'according to the rules'. Contrary to popular belief, a creative *genre* may be based on precisely specifiable rules.

One example (discussed later in this chapter) is music. Western music springs from a search-space defined by the rules of harmony, and its melodies are pathways through a precisely mappable landscape of musical intervals. There are rules about musical tempo, too, which define a metrical search-space. An acceptable melody (a series of notes that is recognizable as a tune) must satisfy both rule-sets – perhaps with some tweaking at the margins. But these rules, though intuitive, are not self-evident. As we shall see, it is no easy matter to discover them, and to show how they make musical appreciation possible.

A domain whose rules are more easily available is chess. Here, the search-space consists of all the board-states that could be reached by any series of legal moves. Each legal move involves a specific action defined by the rules of chess. And each action is constrained by one or more preconditions, without which it cannot be performed. For instance, only on its first move can a pawn advance by two squares, and only when it is capturing an opposing piece can it move diagonally. A precondition may be simple or complex (castling in chess has fairly complex preconditions, and is not taught to beginners). In either case, there must be some effective procedure capable of deciding whether the precondition is satisfied. (Chess-masters get their ideas about promising moves 'intuitively', by perceiving familiar board-patterns; but every move, no matter how it was suggested, must fit the rules.)

Human thought-processes, and the mental spaces they inhabit, are largely hidden from the thinkers themselves. The sort of thinking that involves well-structured constraints can be better understood by comparing it with problem-solving programs, whose conceptual spaces can be precisely mapped. Imprecise thinking – such as poetic imagery, or the intuitive recognition of chess-patterns – can be understood computationally too, as we shall see in Chapter 6. In general, then, AI-concepts help us to think more clearly about conceptual spaces of various kinds.

Some conceptual spaces map more locations, and/or more different sorts of location, than others. A search-space (such as chess) that is defined by many different possible actions, some having complex preconditions, is more highly structured – and therefore richer in its potential – than one (such as noughts and crosses) defined by only a few actions, each having very simple preconditions.

Another way of putting this is to say that the *search-tree* for chess is larger, and more profusely branching, than that for noughts and crosses. The search-tree is the set of all possible action-sequences leading from one legal problem-state to another. The twigs and branches of the tree arise at the choice-points, where the problem-solver must select one of two or more possible actions. In short, the search-space maps the locations, and the search-tree maps the pathways by which they can be visited.

Certain locations may be accessible only from particular starting-points, or only given certain data. Thus in the necklace-game, you needed a particular ready-made necklace to be able to do '2 + 2 = 4'. Only a chemist, who knows which atoms can be neighbours, could discover a new molecular structure. And only someone knowing English vocabulary (as well as grammar) can produce sentences mentioning purple-spotted hedgehogs.

Data and action-rules together comprise a generative system, with the potential (in principle) to generate every location within the conceptual space. The number of these locations may be very large, even infinite. The necklace-game can generate the integers to infinity, and English grammar generates indefinitely many sentence-structures (each of which can be 'filled' by many different sets of words). Chess allows for a number of possible board-positions which, though finite, is astronomical.

In practice, then, not all locations in a large search-space can actually be visited. Moreover, some locations may be irrelevant to the task in hand (the personalized necklaces for the stuffed animals have only *odd* numbers of blue beads on either side of the white one). Even for a small search-space, it may be a waste of time and computational effort to consider every region. In general, one hopes not only to find the right path, but to do so as quickly as possible.

Heuristics are used – by people and programs alike – to prune the search-tree. That is, they save the problem-solver from visiting every choice-point on the tree, by selectively ignoring parts of it.

In effect, they alter the search-space in the mind (or the program). They make some locations easier to reach than they would have been otherwise; and they make others inaccessible, which would have been accessible without them. Some heuristics are guaranteed to solve certain classes of problem. Others can sometimes divert the problem-solver from that part of the search-space in which the solution lies. However, if the heuristic is routinely treated as provisional, or if it can be 'playfully' dropped, then the possibility of visiting the relevant part of the space can be revived.

The early AI problem-solvers of the 1950s used heuristics to solve simple logical problems, and familiar puzzles translated into logical terms. For instance, even these relatively primitive AI-programs could solve the 'missionaries and cannibals' puzzle.

In case you do not know this teaser already, here it is: *Three missionaries and three cannibals are together on one side of a river. They have one rowing-boat, which can hold up to two people. They all know how to row. How can they all reach the other side of the river, given that – for obvious reasons – there must never be more cannibals than missionaries on either river-bank?* I'll leave you to puzzle over this one by yourself. Unlike the necklace-game, it's not easily done with pencil and paper: try using coins, instead. (If you need a clue, think of the heuristic commonly referred to by the French phrase *reculer pour mieux sauter.*)

Since then, countless heuristics have been defined within AI-research, whereby unmanageable tasks have been transformed into feasible ones. Some are very general, others highly specific – to figure-drawing, music or chemistry. And some, as we shall see, have even generated historically new knowledge.

Heuristics are often ordered, those with high priority being applied to the search-tree first. For instance, it is normally advisable to concentrate on substance before tidying up form. If your current state (your location in the problem-space) differs from your goal-state by lacking a content-item, it is probably sensible to give priority to achieving that item before worrying about just how to relate it to other items.

A homely example of this heuristic concerns going on holiday, when it is best to collect all your holiday-things together *before* you start packing your suitcase. Similarly, programs designed to solve logic-problems usually focus first on the substantive terms, leaving their precise logical relations to be adjusted later.

A different ordering of the very same heuristics would imply a different search-tree, a different set of paths through computational space. Human expertise involves knowing not only what specialized heuristics to use, but also in what order to use them.

So a dress-designer, before deciding on the pattern layout for cutting, asks whether the fabric is to be cut on the bias or not. And (as we shall see below) an expert musician, familiar with the conventions of fugue-composition, can immediately restrict an unknown fugue to only four possible home-keys on hearing the very first note. In other words, if the relevant musical heuristic is applied immediately, rather than at some later stage in the interpretative process, it narrows the field so much that a host of potentially relevant questions need not even be asked.

In dealing with unusual problems, flexibility is required. For example, fugues occasionally begin in a way that defies the usual conventions. To interpret these 'rogue' fugues successfully, the musician must treat the initial four-key restriction as merely *provisional*. An experienced musician may do this routinely, having learnt that composers occasionally ignore – or deliberately break –this particular rule of fugue. But suppose a less experienced musician encounters a rogue fugue, for the first time. The novice could deal with it P-creatively (involving psychological, as opposed to historical, novelty), by using the general heuristic of *dropping a heuristic* to break the grip of the relevant rule.

This is sometimes easier said than done. Human experts used to thinking – painting, composing, doing chemistry – in a certain way may be unable to capitalize fully on their own mental resources, because certain habits of thought cannot be overcome. A heuristic that cannot be dropped, or even postponed, may be very useful in normal circumstances. But when normal (P-creative) thinking, within a different conceptual space, is required the frozen heuristic can prevent it.

We have seen that a heuristic can sometimes be dropped, so putting previously inaccessible parts of the search-space back on to the map. But heuristics can be *modified*, too. A problem-solving system – program or person – may possess higher-order heuristics with which to transform lower-level heuristics.

'Consider the negative', for instance, can be applied not only to individual problem-constraints (turning *strings* into *rings*, as described in Chapter 4) but also to other heuristics (now positively suggesting the sacrifice of the queen). In either case, a new – and perhaps a radically different – search-space is generated, which nonetheless shares some features with the previous one. Some AI-work to be discussed in Chapter 8 is specifically focused on *heuristics for changing heuristics*, so that computer problem-solving might be able, like ours, to produce a fundamentally different conceptual space.

By a 'fundamental' difference in conceptual space, I mean one due to a change at a relatively deep level of the generative system concerned.

Many generative systems define a hierarchical structure, some rules being more basic than others. Represented as a search-tree, the basic choice-points occur at the origins of the thickest branches, while the more superficial choices give rise to the twigs. To chop off one branch (to drop a fundamental constraint) is to lose all the twigs sprouting from it – which may be a significant proportion of the whole tree.

Consider English grammar, for instance. Every sentence must have a noun-phrase and a verb-phrase. The noun-phrase, in turn, may (though it need not) contain one or more adjectives. So these are all acceptable sentences: 'The cat sat on the mat', 'The black cat sat on the mat,' 'The sleek black cat sat on the mat.'

If some all-powerful tyrant were to issue an edict that no more than one adjective could be attached to any noun, the last sentence would be forbidden (and Dickens could not have written 'a squeezing, wrenching, grasping, scraping, clutching, covetous old sinner' without risking punishment as well as admiration). But this neo-English would still be recognizable, and intelligible, as a sort of English. Remove the rule saying that a noun-phrase may contain an adjective, and we would have a still more simplified 'English', in which many things could be said – if at all – only in a very different fashion ('The cat has sleekness'). And remove the most basic grammatical constraint of all, that there be a noun-phrase and a verb-phrase, and a random string of English words (like *priest conspiration sprug harlequin sousewife connaturality*) would be no less admissible than 'The cat sat on the mat.'

Similar remarks apply to music. Modulations from one key to another, however unusual and/or closely juxtaposed they may be, are less fundamentally destructive of tonality than is ignoring the home-key altogether.

These examples suggest an answer to the question (posed at the end of Chapter 3) how one impossible idea can be 'more surprising' than another. The deeper the change in the generative system, the more different – and less immediately intelligible – is the corresponding conceptual space.

In some cases, the difference is so great that we speak of a new sort (not merely a new branch) of art, or science, and assign a greater degree of creativity to the innovator. To add rings to strings, as a class of molecular structures, is less creative than thinking of string-molecules in the first place. And to think of strings within the conceptual space of (Daltonian) chemistry involves a less radical change than to pass from the mediaeval elements of fire, air, earth and water to the elemental atoms of modern chemistry. Accordingly, John Dalton (who developed the atomic theory) is a more important figure than Kekulé in the history of chemistry.

People often claim that talk of 'rules' and 'constraints' – especially in the context of computer programs – must be irrelevant to creativity, which is an expression of human freedom. But far from being the antithesis of creativity, constraints on thinking are what make it possible.

Constraints map out a territory of structural possibilities which can then be explored, and perhaps transformed to give another one. Dickens could not have created his luxuriant description of Scrooge without accepting the grammatical rule about adjectives, and pushing it towards its limits. If the child had been allowed to build *just any* necklace on that wet Sunday afternoon, she could never have done addition-by-necklace and would never have had the idea of doing subtraction-by-necklace. (A necklace might happen to have equal numbers of blue beads on either side of the white one; but 'addition' would not be possible because, since anything is admissible, this property need not apply to its descendants.)

Similarly, it is no accident that Schoenberg, having abandoned the constraints of tonality, successively introduced others – using every note of the chromatic scale, for instance. Whether his added constraints are aesthetically pleasing, as opposed to being merely ingeniously productive, is another question. Some people would argue that they are not, because they are arbitrary with respect to the natural properties of auditory perception. (We shall see later that some artistic genres are not arbitrary in this way. Impressionism, for instance, exploits deep properties of vision; despite the Impressionists' concern with the science of optics, their work is therefore less 'intellectual' than Schoenberg's music.)

In short, to drop all current constraints and refrain from providing new ones is to invite not creativity, but confusion. (As for human freedom, I shall suggest in Chapter 11 that this, too, can be understood in computational terms.)

This does not mean that the creative mind is constrained to do only one thing. Even someone who accepts all the current constraints without modification will have a choice at certain points – sometimes, a random choice would do. Bach was constrained, by his own creative decision, to compose a Fugue for the 'Forty-Eight' in the key of C minor; and that meant that he was constrained to do certain things and not others. But *within* those musical constraints, he was free to compose an indefinitely large range of themes. Likewise, to speak grammatically is not to be forced to say only one thing.

When all the relevant constraints have been satisfied, anything goes. And 'anything' can include idiosyncratic choices rooted in the creator's personal history, or even purely random choices based on the tossing of a coin.

It is the partial continuity of constraints which enables a new idea to be recognized, by author and audience alike, as a creative contribution. The new conceptual space may provide a fresh way of viewing the task-domain and signposting interesting pathways that were invisible – indeed, impossible – before.

Thus Kekulé's novel suggestion about a single molecule, benzene, implicitly created an extensive new search-space, whose locations (ring-structures) had been chemically inconceivable beforehand. But many of the preconditions already accepted in chemistry remained. These included constraints (such as valency) on which atoms could be linked to which, and the necessity of fitting the theory to the experimental data.

Similarly, tonality makes a whole range of musical compositions possible, and successive refinements and/or modifications of it generate new possibilities in turn. But until the final break into a fully chromatic search-space, the persisting musical conventions – about the return to a home key, or preferred modulations or cadences – provide the listener with familiar bearings with which to navigate the unfamiliar territory. Even the flight into atonality can be understood as the final step in a progressive structural modification of tonal space, as we have seen.

This explains our reluctance (noted in Chapter 3) to credit someone with an H-creative, or historically original, idea who merely had the idea – without recognizing its significance. To come up with the notion of elliptical orbits only to reject it – as both Copernicus and, initially, Kepler did – exemplifies cosmic irony rather than astronomical creativity.

You may think it unjust, not to say impertinent, to make such a remark about these two H-creative geniuses. You may prefer to say that their thinking about elliptical orbits did not achieve 'full-blown' creativity, or even that it was 'creative but unsuccessful'.

Such ways of putting it remind us that the rejected idea was neither random nor perverse (as it might have been had it arisen in the mind of an uneducated crossing-sweeper or a semi-educated crank), but arose while exploring the relevant conceptual space in an intelligible way. They remind us, too, that the idea was correct; it satisfies the evaluative aspect of creativity. The point, however, is that this evaluation was not made by Copernicus nor (at first) by Kepler, who called his novel idea 'a cartload of dung'.

Unless someone realizes the structure which old and new spaces have in common, the new idea cannot be seen as the solution to the old problem. Without some appreciation of shared constraints, it cannot even be seen as the solution to a new problem intelligibly connected with the previous one. This is why original ideas – even when valued by their originators – are so often resisted, being accepted only by a handful of like-minded *aficionados*.

It is the constraints, whether implicitly understood or explicitly recognized, which underlie the 'expectations' mentioned (when defining creativity) in Chapter 3. The more expectations are disappointed, the more difficult it is to see the link between old and new.

This is not a question merely of counting expectations: it also involves assessing their generative depth (their point of origin on the search-tree).

The music-lover can accept – even appreciate – an unusual modulation, a surprisingly dissonant chord, within a recognizably tonal context. But drop the home-key, and almost all familiar bearings are lost: the old map is destroyed and it is not obvious how to construct a new one. Similarly, the word-string *priest conspiration sprug harlequin sousewife connaturality* fails to satisfy even our deepest expectations about grammatical structure, so is unintelligible and usually worthless (except as a stimulus to 'free associations' in thinking). James Joyce might have made something of it, as remarked in an earlier chapter, but only by setting up a novel context of expectations. And even he could not have got away with the total eclipse of grammar.

Creative ideas are surprising, yes. They go against our expectations. But something wholly unconnected with the familiar arouses not surprise so much as bewilderment. To be sure, the lack of connection with what went before may be apparent rather than real. But someone to whom the connection is not apparent will not be able to recognize the idea as *creative* (as opposed to *new*). Nor will they be able to see it as relevant to what they had regarded as the problem-domain in question: 'That's not art!', 'Call that poetry?', 'It's a tissue of [chemical] fancies!'

You may object: 'Constraints, yes. Computer programs – never!' But since creativity is a question of what thoughts *can* and *cannot* result from particular mental structures and processes, anyone seeking to understand it needs to be able to describe those structures and processes clearly, and assess their generative potential rigorously. This is why it is helpful to use AI-terms to describe the creative constraints in human minds.

For AI-concepts must be unambiguously defined, if they are to be embodied in a computer program. Moreover, any result obtained when actually running a program *must* (ignoring hardware faults) lie within the potential of the program concerned.

This practical test of a generative system's potential is crucial. In principle, to be sure, it is dispensable. A computer can do only what its program, and data, enables it to do. In Lady Lovelace's words, it can do only 'whatever we know how to order it to perform'. So someone (God?) with perfect memory and enormous computational power could assess the generative potential of any computer program without actually running it on a computer. To a limited extent, human computer scientists can do the same thing (much as you could recognize the necklace-game's power to generate all the integers). But although God could never be surprised, human programmers often are.

No matter how great our surprise, however, the fact that a program does something is conclusive proof that it has the generative power to do so. The structural and procedural constraints embodied in it are, without any doubt, rich enough to make such computations possible.

To see how a computational psychology can help in identifying the creative constraints within our minds, let us take music as an example.

Consider the tacit interpretation that you carry out, when you hear an unknown melody (from Western music) and 'intuitively' recognize its metre and key. On first hearing the melody, you can usually start beating time to it very soon (you can perceive its metre very quickly). And if the singer or instrumentalist plays a wrong note, you can wince at the appropriate moment – even though you have never heard the tune before, and may not know just what the right note is.

If you are musically trained, your interpretation of metre and harmony can be made publicly visible. For you will usually be able to write the melody in musical notation – maybe humming it through several times while doing so. (If you are not musically trained, some terms used in the following pages will be unfamiliar. This does not matter: the points that are relevant for our purposes are very general, and can be grasped without a specialist knowledge of music.)

Assuming that you lack the very rare gift of perfect pitch, the first note must be identified for you. This can be done by someone touching a particular black or white note on a piano keyboard. (Touching a note gives you no clue to the home-key, whereas naming it – as F-sharp, or G-flat, or E-double-sharp – does.)

You can then do your exercise in 'musical dictation'. Probably, you will be able to specify the time-signature, bar-lines, note-lengths, rests, key-signature and notes. Also, you will be able to identify any accidentals: sharps, flats or naturals. (Let us assume that the melody is sung or played dead-pan, so you need no marks of expression like *staccato*, *legato*, *crescendo* or *rallentando*.)

Moreover, if someone offers you two alternative transcriptions of the melody, each picking out exactly the same notes on the keyboard, with exactly the same durations, you will be able to see that one is right and the other wrong. For instance, starting with a middle-A crotchet, one could write 'God Save the Queen' either in 4/4 time in the key of B-flat major, or in 3/4 time in the key of A major (see Figures 5.1(a) and (b)). It is intuitively obvious, even to a musical novice, that the second form is correct and the first is 'crazy'.

Figure 5.1 (a)

Figure 5.1 (b)

That is, we can resolve grammatical ambiguity in music, much as we do in language. Sometimes, we change our interpretation of a heard melody as it proceeds. (Compare the infant hymn-singer's realization that the line she has been lisping fervently for months must be 'Gladly my cross I'd bear', not '*Gladly*, my cross-eyed bear.') But our interpretation usually stabilizes well before the melody is finished.

How are these musical responses possible? How can someone not only recognize an unfamiliar series of sounds as a *melody* (compare: a *sentence*), but even locate it accurately within musical space? What mental processes are involved, and just what knowledge of musical structure is required? And how can one written version be better than another, if the sounds represented are identical? In short, what maps of musical space, and what methods of map-reading, are used in perceiving melodies?

These questions must be answered, if we are to understand musical creativity. For the appreciation of an idea requires some of the very same psychological processes that are needed to produce it. A person requires a map of musical space not only to explore the space, or transform it, but also to locate unfamiliar compositions within it. (There is nothing unique about music in this; our 'intuitive' grasp of English, for instance, requires sensitivity to grammatical structure.)

Christopher Longuet-Higgins (a fine musician, and a computational psychologist too) has discussed these matters with great sensitivity – and clarity.[1] He provides a theory of the structure of harmonic (and metrical) space, and of the mental processes by which we interpret (single-voiced) melodies.

His papers are computational, but they are not about computers. (In his two 'letters to a musical friend', computers are not even mentioned.) On the contrary, they are about music, and how it is possible for people to appreciate it. They can deepen our understanding of the creative developments in tonality mentioned in Chapter 4. For his map of harmonic space is not 'flat'. It distinguishes more and less fundamental features of this conceptual landscape, and helps to show which avenues of exploration are the most adventurous, the most likely to lead to really shocking surprises.

Consider harmony, for example. Harmony concerns not notes (pitch), but the relations between notes. A theory of harmony should describe the intervals that can occur in tonal music, and show how they are related. Also, it should explain why modulations from one key to another follow certain paths, some of which seem more 'natural', and were explored earlier, than others.

Longuet-Higgins shows that *every interval that can possibly occur* in tonal music can be expressed, in one and only one way, as a combination of octaves, perfect fifths and major thirds. In other words, he depicts tonality as a three-dimensional space, structured by these three basic intervals. (Previous theories of harmony, since Helmholtz, had referred only to octaves and perfect fifths.) A composer choosing the 'next note' of a tonal melody *must* choose one that is related to its predecessor by some specific interval selected from this search-space. Moreover, the theory defines the mutual relations between intervals, some interval-pairs being closer within harmonic space and others more distant.

A key corresponds to a specific region within tonal space. When defining a key, one can ignore octaves. In the key of C major, for instance, 'middle C' and 'top C' – despite their difference in pitch – play the same harmonic role; so do the two Ds, adjacent to them. For the purpose of defining keys, then, we can drop the octave-dimension and treat tonality as a two-dimensional space. Every harmonic interval *within a given key* is definable in terms of perfect fifths and major thirds.

Because there are now only two musical dimensions to consider, harmonic space can be represented by a very simple (two-dimensional) diagram. Longuet-Higgins constructs a spatial array, in which each note is one perfect fifth higher (in pitch) than the note on its left, and a major third higher than the note written underneath it (Figure 5.2).

A♯	E♯	B♯	Fx	Cx	Gx
F♯	C♯	G♯	D♯	A♯	E♯
D	A	E	B	F♯	C♯
B♭	F	C	G	D	A
G♭	D♭	A♭	E♭	B♭	F
E♭♭	B♭♭	F♭	C♭	G♭	D♭

Figure 5.2

If we mark, on this array, the notes occurring within any given key, or scale, we find that they occur in clusters of neighbouring notes. Moreover, these clusters have different shapes, according to whether the key is major or minor. Harmonically equivalent notes (tonic, dominant, sub-mediant and so on) have the same relative position within any major (or minor) cluster. For example, the tonic of a major scale is second from the left on the bottom row, and the tonic of a minor scale is the second from the left on the top row (see Figure 5.3, in which the boxes show C major and D minor).

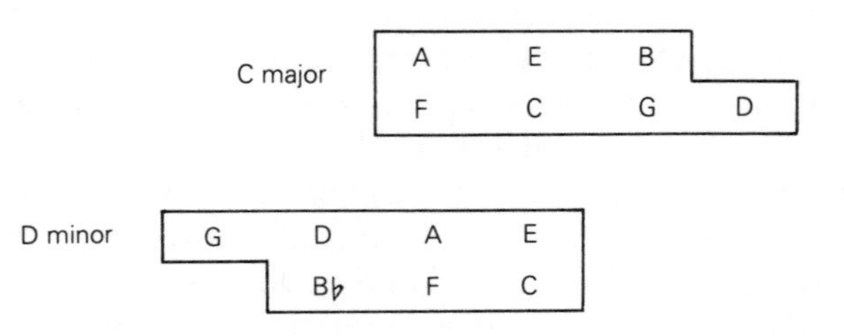

Figure 5.3

As you can see, the key-boxes overlap each other. Modulations from one key to another exploit the fact that any two keys will share at least one note (C major and D minor share four). The specific pathways, or sequences of transitions, by which the composer can modulate between keys are mapped within tonal space, some being more direct than others. The musical subtleties involved are considerable, but Longuet-Higgins shows in detail (which need not concern us) how they follow from his theoretical analysis.

Longuet-Higgins calls his theory of tonal intervals a 'grammar' of harmony. (He also defines a metrical 'grammar', which describes the rhythmic structure of most Western music.) Like English grammar, it allows for certain structures but not others.

However, grammatical structure is one thing: procedures for parsing it are another. (And procedures for composing grammatical structures in the first place are different again.) To map the underlying harmonic structure of all melodies is not to say what methods of map-reading will enable us, as listeners, to find the harmonies in a specific melody. But parsing must clearly be sensitive to grammar. Whatever procedures we rely on to interpret harmony (or metre), they must exploit the structural regularities involved.

Accordingly, Longuet-Higgins uses his theories of harmony and metre in defining the map-reading methods described in his papers. These are musical heuristics, ordered for priority, on which we may rely when we 'intuitively' perceive the harmony and rhythm of melodies. (Other perceptual processes are doubtless involved as well; and in polyphonic music, harmonic information is available in the chords.) They suggest exactly what questions the listener (unconsciously) asks, and they exhibit the logical relations between the various answers. In short, they specify multiple paths through a search-tree for assessing the key or tempo of a melody.

They include, for instance, a 'rule of congruence'. This rule applies to both harmony and metre, and explains why we can assign the key to a melody, and start beating time to it, long before the end. It states that, until the correct key or time-signature has been established, no non-congruent note can occur (unless the note is non-congruent with all possible signatures). So no accidentals can occur before the key is established. (Further rules are given which state how one can decide whether or not a key has been established.)

Perhaps you can think of counter-examples. But if you turn to the detailed analyses given in Longuet-Higgins' papers, you will probably find that these are covered by the subsidiary rules given there. Occasionally, your counter-example will stand. However, if you are a good enough musician to have thought of it in the first place, you may sense that the composer was deliberately 'playing' with the listener by creatively ignoring this near-universal constraint.

The rule of congruence is very general. Some of the other heuristics are less so. Instead of applying to all tonal music, they apply only to a certain type.

For instance, it is a standard rule of fugue (occasionally broken) that the first note is either the tonic or the dominant. That is why, as remarked above, the expert musician who knows that a piece is a fugue can eliminate almost all key-signatures on hearing the very first note. However, to include the rule in that stark form in explaining our perception of fugues would be, in effect, to cheat. After all, the musical amateur (who does not know this rule) can identify the key as the melody proceeds. So Longuet-Higgins defines a rule of 'tonic-dominant preference', to be used only as a last resort. (It states that if the listener is in a dilemma arising in one of two precisely defined ways, then the first preference is a key whose tonic is the first note, and the second preference is one whose dominant is the first note.) As he intended, this rule is only rarely needed. Many fugues can be correctly transcribed without using it.

Some heuristics are even more specific, applying to a certain composer's style. One concerns Bach's use of chromatic scales, and helps to find the accidentals in his fugues. (Possibly, this rule would be useful with respect to some other composers, too. The degree of generality of such heuristics is an interesting research area for musicology.)

The map of tonal harmony drawn for us by Longuet-Higgins was not spawned by arbitrary speculations, but derived by abstract argument from first principles. And he applies it to many examples of post-Renaissance music – citing passages from Handel, Bach, Purcell, Mozart, Beethoven, Brahms, Schubert, Chopin, Wagner, Elgar and others.

Nevertheless, sceptical readers might feel that they have only his word (and their own shaky musical intuitions) to go on. If these matters escaped even Helmholtz, one may be forgiven for doubting whether this new view of harmony is correct. And even if it does describe abstract harmonic structure, can it really explain the *process* of harmonic interpretation by the listener – the successful recognition of key, accidentals and all? In other words, can this theory really chart the pathways by which we find our way in musical space?

This is where functioning programs come in. Programs based on Longuet-Higgins' theory have succeeded in interpreting (transcribing) the fugue subjects in Bach's 'Well-Tempered Clavier' and melodies from Sousa and Wagner, and have been applied also to jazz. Provisional assignments of key and metre are made almost at the start of the piece, becoming more definite, and sometimes different, as the melody continues. Usually, they stabilize well before the end, as our own musical intuitions do. (Details of the heuristics, and the program-code, are given in the relevant papers.)

Longuet-Higgins' theories of musical structure, and of the mental processes involved in perceiving it, are highly general. His programs are widely applicable too, for they can transcribe many different tonal compositions. One of his papers was initially rejected by the sceptical editor of *Nature*, on the grounds that coaxing a computer to transcribe the march 'Colonel Bogey' was too trivial a task to be interesting. The suggestion was made – seriously? sarcastically? – that he use something from Wagner instead: if the program managed to handle that, they would publish his paper. He did, it did and they did.

Despite this story-with-a-happy-ending, it must be admitted that his programs have many limitations. *A fortiori*, they cannot interpret (compare: appreciate or understand) those cases where the composer 'breaks the rules'. Longuet-Higgins points out, for instance, that the program which finds the correct key-signatures for all the fugues of the 'Forty-Eight' would assign the wrong key to one of the fugues in the 'Mass in B Minor', because Bach there ignored a constraint (the rule of congruence) that he usually honoured. But the constraint concerned is precisely specified within Longuet-Higgins' computational approach – which is why one can identify Bach's creativity in ignoring it.

His computational approach helps him, too, to notice and eliminate weaknesses in his theories. Some melodies are incorrectly transcribed because of inadequacies in the programmed rules – although, to Longuet-Higgins' credit, some of the mistakes are ones which a human musician might make too. (To his credit, because a psychological theory should explain people's failures as well as their successes.) However, these inadequacies can be clearly identified, so spurring further study of the conceptual spaces concerned.

For example, the early metrical program let loose on Bach's 'Well-Tempered' fugues could not deal with the (six) cases where all the notes and rests are of equal duration. The reason was that the program had to compare notes (or rests) of unequal length in order to work out the time-signature. Presumably, human listeners can use 'expressive' temporal features (such as accent, slurring, *rubato*, *legato* and *staccato*) as additional clues.

But how? Just what computational processes are involved when one uses phrasing to help find the tempo? In a recent paper, Longuet-Higgins has extended his theory of metrical perception (originally consisting of rules that rely only on information about the times of *onset* of notes) to include a phrasing-rule based on slurring, where the 'offset' of one note coincides with the onset of the next. This enables one to parse the metrical structure of melodies having all-equal notes and rests, and even of music that includes syncopated passages.

More recently still, he has formulated rules which enable a computer to play Chopin's 'Minute Waltz' in an expressive, and 'natural', manner. The dead-pan performance, played without the benefit of these rules, sounds musically dead – even absurd. But with these further constraints added, the computer produces a performance that many human pianists would envy.

What do these music-programs have to offer to someone who wants to understand the human mind? Two things.

First, they identify a number of clearly stated hypotheses which psychologists can investigate. For example, they suggest that our perception of rhythm and of harmony involve independent processes, and that our sense of phrasing depends on melodic 'gaps' defined in one way rather than another, more obvious way.

Second, they prove that it is *possible* for a computational system to use Longuet-Higgins' musical grammar and interpretative procedures for correctly transcribing a wide range of melodies. Since his programs – like all programs – are effective procedures, there can be no doubt about this point.

Whether people perceive melodies by means of mental processes exactly like the heuristics used in Longuet-Higgins' programs is another question. Whatever program we consider, it is always possible that a different one might produce the same results. This abstract possibility must be weighed against the practical difficulty of coming up with even one successful program, even one psychological theory that fits the facts. But a program may have specific features that are psychologically implausible, along with others that are more credible.

One difference between how we interpret music and how Longuet-Higgins' programs do so concerns *serial* as opposed to *parallel* processing. It is most unlikely that the human mind interprets melodies by asking the relevant questions one by one, in a strict temporal order. More probably, it asks them in parallel (functioning as a neural network or 'connectionist' system, as described in the next chapter).

That is, it seeks multiple harmonic (or metrical) constraints simultaneously, arriving at a stable interpretation of the melody when these constraints have been satisfied in a mutually consistent way.

But if Longuet-Higgins' theory of musical perception is correct, the network must (in effect) ask the questions he identifies. Similarly, it must respect the logical relations that he exhibits between the various possible answers. Something which is musically inconsistent must be recognized as such by the network.

In short, heuristics can be embodied in parallel-processing systems. It is a mistake to think that sequential computer programs cannot possibly teach us anything about psychology. 'Search-tree' theories may identify some of the specific computational processes which, in human beings, are run in parallel. This point applies to all domains, not just music. There is great excitement at present about the recent AI-work on connectionism. This is understandable, as we shall soon see. But it should not obscure the fact that step-by-step AI-models, despite their 'unnatural' air, can help us investigate the contents, structures and processes of human thought.

A second difference between Longuet-Higgins' computer programs and musicians' minds concerns the way in which harmonic relations are represented within the system. The program which found every key sign, and notated every accidental, in Bach's 'Forty-Eight' fugue subjects used a most 'unnatural' representation of the basic intervals and keys. No one imagines that human listeners refer, even unconsciously, to an inner spatial array carrying variously shaped boxes. The program does not model our perception in *that* sense.

However, we saw in Chapter 3 that a generative system can be viewed either as a set of timeless descriptive constraints or as a specification of actual computational processes. It follows that the basic grammar of tonal harmony may be captured by Longuet-Higgins' musical theory, even though human minds do not use arrays and boxes to parse it.

Moreover, this theory tells us that *whatever* the processes are by which we manage to parse melodies, they must give perfect fifths and major thirds a fundamental place in the interpretative process. The theory also suggests various specific heuristics, or interpretative rules, as we have seen. The success of the program supports such psychological hypotheses, even though it uses a spatial array to represent the harmonic dimensions concerned whereas human minds do not.

Or rather, most human minds, most of the time, do not. But Longuet-Higgins and his readers – and now you, too – can use the array-and-boxes representation as an external aid in thinking about harmony. Indeed, one aspect of Longuet-Higgins' own creativity was his coming up with this particular *representation* of his harmonic theory. In general, problem-solving is critically affected by the representation of the problem that is used by the problem-solver.

Many creative ideas enable us to think about a problem in a new way by representing it in terms of a familiar analogy. For instance, Niels Bohr's early representation of the atom as a solar system made possible fruitful questions about the numbers and orbits of electrons, and modulations (quantum jumps) between one orbit and another. Similarly, William Harvey's description of the heart and blood-vessels as a hydraulic system enabled him to explain – and discover – many different facts about the blood-supply.

Other original ideas modify an existing representation. For instance, Kekulé's intuitive leap from strings to rings extended the representational system then used in chemistry.

Some of the most important human creations have been new representational *systems*. These include formal notations, such as Arabic numerals (not forgetting zero), chemical formulae, or the staves, minims and crotchets used by musicians. Programming languages are a more recent example, being notational systems that make it possible to state – and to develop – effective procedures of many different kinds.

As well as recording new ideas for posterity, such notations may make them possible in the first place. For a written language can help us to explore the implications of the ideas it represents. It enables us to write down, and so remember, our previous thoughts – including the passing-thoughts, or 'intermediate results', crucially involved in reaching a conclusion.

An adequate explanation of creativity should include a systematic theory explaining how, and why, different sorts of representation are appropriate to different classes of problem. As yet, no such theory exists. But representation, and the difference it makes to problem-solving, is much studied in AI.

Various methods of representing knowledge have been used in computer programs, including *scripts*, *frames* and *semantic nets* – sometimes called *associative nets*. (Each of these concepts will be explained soon.) In addition, some theoretical AI-research has tried to distinguish general types of representation; an example discussed below is *analogical* representation, of which Longuet-Higgins' array-and-boxes is a special case.

The concept of *semantic net* originated in psychology. It depicts human memory as an associative system wherein each idea can lead to many other relevant ideas – and even to 'irrelevant' ideas, linked to the first by mere coincidence. For instance, *violet* may call to mind not only *colour* and *flower*, but also *living*, *woodland* and *springtime* – and even (for someone like myself, whose mother's name was Violet) *mother*; each of these ideas has further associations (*springtime* with *Paris*, for example), and so on.

Semantic nets in AI are computational structures representing (in a highly simplified way) the field of meaning within a certain part of conceptual space. They are often used not for doing 'logical' problem-solving, but for modelling spontaneous conceptual associations.

A semantic net consists of nodes and links. The nodes stand for specific ideas, while the links – whereby one idea can be accessed from another – represent various types of mental connection.

Most links have some semantic relevance, although it is possible also to include links coding 'meaningless' and idiosyncratic associations (like that between *violet* and *mother*). Semantically significant links can represent not only specific properties (connecting *violet* and *sweet-smelling*), but also structural matters such as *class-membership*, *similarity*, *instantiation* and *part–whole* relations. (A violet is a member of the class of flowers; cats are similar to dogs, in being four-legged domesticated animals; a red setter is an instance of 'dog'; and a finger is part of a hand, which is part of a body.)

The structure of the semantic net may enable 'spontaneous' inferences to be made by means of pre-existing links. For example, a node can inherit properties from a more inclusive class (if flowers are a form of life, and violets are flowers, then violets too are alive). Some links can hold only between ideas of certain types: *friend of* can connect people, but not pots and pans (except in a fairy-story, wherein pots and pans can speak and hope and argue, too). A programmed semantic net may contain only one sort of link, or several; the tiny net shown in Figure 5.4, for example, has six types.

The 'meaning' of an idea represented within a semantic net is a function of its place in the system. It involves not only the node specifically labelled for the idea in question (*violet*, perhaps) but also all the nodes that can be reached, directly or indirectly, from that node. Processing may proceed through the net until all the relevant pathways have been pursued to their end-point (or returned to their origin). Alternatively, journeys may be restricted to a maximum number of links. If so, then increasing the number specified in the cut-off rule will make paths available which were impassable before.

If potential pathways exist connecting any and every node within the

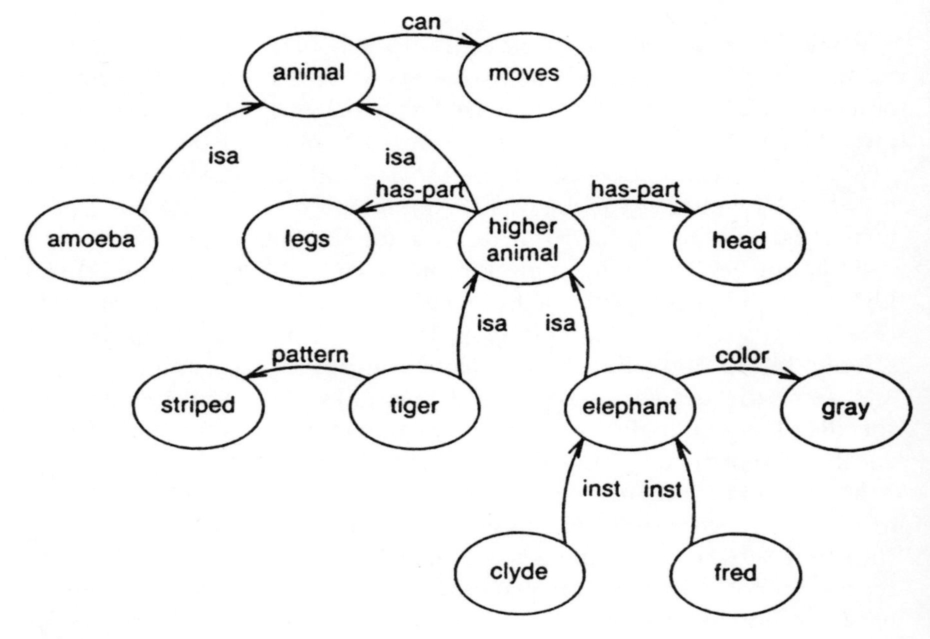

Figure 5.4

network, then the meaning of each individual node is a function of all the others. In this case, adding a new node will implicitly affect the meaning of all the existing ones. Analogously, a single experience – or poetic image – may subtly change the significance of a large range of related ideas within someone's mind. In other words, new meanings (or shades of meaning) are created which were not possible before. Having heard Hopkins' description of thrushes' nests as 'little blue heavens', one thinks of them in a new way.

Many writers see the creative, largely unconscious association of ideas as being primed by the 'preparatory' phase of conscious work. They assume that if the latent ideas can be *selectively* activated, there is a fair chance that some of the resultant associations will be relevant to the creator's concerns.

The concept of semantic nets suggests how this selective (but largely unconscious) preparation can happen. From a selected node, or set of nodes, it may be possible – especially if some of the 'sensible' cut-off rules are temporarily weakened – to reach other nodes that would otherwise be inaccessible. In Chapter 7, for instance, we shall discuss a semantic net containing over thirty thousand words, and many hundreds of thousands of links, which enables concepts to be compared in many different *and specifiable* ways.

Scripts and *frames* can be thought of as special cases of semantic nets. They represent the conceptual skeleton of a familiar idea – such as how one should behave in a restaurant, or what it is to be a room, or a molecule. In other words, they are schematic outlines of a tiny part of conceptual space. (Like semantic nets in general, they can be modelled by connectionist systems – in which case, they show the conceptual tolerance, or 'fuzziness', to be described in Chapter 6.)

Scripts concern socially recognized ways of behaving (involving the roles of customer and waiter, for instance), whereas frames usually represent single concepts or ideas. Although the distinction between scripts and frames is not clear-cut, their equivalents in human minds are likely to be more prominent in literary or scientific creativity, respectively.

One example of a script is Roger Schank's restaurant-script, which is embodied in computer programs that answer questions about stories featuring restaurants.[2] Schank claims that similar computational structures exist in human minds, being accessed whenever we enter or think about restaurants. The restaurant-script is a schematic representation of the behaviour of customers and staff in a hamburger-bar in the United States (it does not tell Parisians just how to behave in Maxim's).

It specifies not only who does what when all goes well, but also how to deal with the unexpected (no menu on the table) or the unacceptable (a burnt hamburger). If the menu is missing, the diner can either request one from the staff-member who seats the customers at their tables, or borrow one from another customer. And if the hamburger turns out to be burnt, the diner normally complains to the waiter or manager and refuses to hand over any money to the payclerk. These specifications of what to do in deviant circumstances are called *what-ifs*, and are an integral part of the script.

Some sorts of deviance are even more unexpected than missing menus and burnt hamburgers, so cannot be anticipated by specially crafted what-ifs. In the absence of ready-made contingency-plans, one has to plan from scratch. *Planning* (mentioned later in this chapter) is an important computational concept, on which a great deal of AI-work has been done. We shall see in Chapter 7 that some programs (designed to write and/or interpret stories) can plan afresh if there are no relevant what-ifs available. For this to be possible, the script must be embedded within a planning-program with access to more general knowledge about means–ends relationships and about what can be done in social situations.

The questions answered by Schank's programs cannot be dealt with by a simple memory 'look-up', because the answers were not explicitly mentioned in the story. (The story may say, 'When Mary's hamburger arrived, it was burnt. She left, and had an Indian meal instead': the script-program can tell us that Mary did not pay for the hamburger, but the story did not actually say so.) Indeed, some question–answer pairs concern events which might have happened, but didn't. The program then has to construct a 'ghost-path' through the conceptual space defined by the script, a sequence of actions which *could* have been followed by some character in the story, but wasn't.

It is because our minds contain something like restaurant-scripts that Jean-Paul Sartre could cite a waiter as an example of 'bad faith', in which one unthinkingly plays a role rather than taking responsibility for making one's own decisions. Acting in bad faith can sometimes be justly criticized. But we could not dispense with scripts entirely. The computational overload involved in making all our decisions about interpersonal behaviour from scratch would be utterly insupportable.

Novelists and dramatists rely heavily on the scripts in the minds of their audience. For to understand even the simplest story, one must implicitly fill many gaps in the action as explicitly reported. Moreover, without script-based expectations the audience could not be surprised by some creative twist wherein a character does something strange. Alice's assumptions about normal human behaviour, not to mention her nice sense of the Victorian proprieties, were continually challenged in Wonderland.

Sometimes, one and the same underlying literary motif is expressed in terms of significantly different scripts. In Chapter 3, we asked whether Shakespeare was 'really' creative in writing *Romeo and Juliet*, whose plot was based on a story by Bandello. One might ask the same thing about Ernest Lehmann and Jerome Robbins, the authors of *West Side Story*. Ignoring the quality of the language involved, one can identify the same motivational themes in either case. Love, loyalty and betrayal: all these play a similar part. A colleague of Schank's has outlined a systematic analysis of these psychological concepts (as we shall see in Chapter 7). But, at this level of abstraction, we have only a plot: not a story.

To convert a plot into a story, we must situate it in some specific time and place. To do this, we must choose (or, in science-fiction, invent – and convey) a relevant set of culture-specific scripts.

Just as Schank's restaurant-script is not perfectly suited to Maxim's, so scripts involving duels with fencing-foils are not well suited to twentieth-century New York. But provided that fencing-foils and flick-knives have the same function in the plot, the one can be substituted for the other. Part of Lehmann and Robbins' creativity was to map the conceptual space defined by Shakespeare on to a new, though fundamentally similar, space defined by the scripts followed by certain groups of present-day New Yorkers.

Frames, as defined in AI, are hierarchical, for they can include lower-level frames – as 'vertebrate' includes 'bird', or 'house' includes 'room'. Various *slots* are defined within the frame, and specific instances of the general class in question are represented by having different details in the (low-level) slots in the frame.

A programmed frame will include suggestions for moving around within the relevant part of conceptual space. That is, it provides 'hints', or computational pointers, suggesting which property or unfilled slot should be considered at various stages of thinking. Some slots come provided with 'default values', so that the program assumes, in the absence of information to the contrary, that the slot carries a particular description.

Human problem-solvers appear to make similar assumptions. For example, someone who sees a room *as* a room, or who recognizes a painting as a painting of a room, or who plans to redecorate an existing room or design a new one, relies on some ideas of what rooms in general look like. A typical room has a roof (not the open sky), a flat floor (not a curved bowl or a swimming-pool), four walls (not six), one door in a wall (not a trap-door in the floor), and one or more windows in the walls (unlike 'internal' bathrooms).

Both frames and scripts are sketch-maps of much visited spots in conceptual space. That is, they represent conceptual stereotypes. It might seem, then, that they can have nothing to do with creativity. But, as Koestler recognized in speaking of the 'bisociation of matrices', novel associations may be mediated by similarities between commonplace ideas.

People can associate frames or scripts with great subtlety, but current AI-technology is more limited. To think *in the way outlined above* about what a room is, and how it relates to a house, is not to recall that 'In my father's house are many mansions'. Similarly, to decide *in the way outlined above* what to do next in a restaurant is not to be reminded of 'A book of verse, a flask of wine, and thou'. A computational system – such as a human mind – that can move from 'room' or 'restaurant' to some distant concept or literary text needs associative links that are more far-reaching, and less tightly constrained, than the inferential procedures typical of frames and scripts. But given a conceptual space in which a rich semantic network links one concept to another, one can suggest how such reminding might be possible.[3]

Moreover, familiar ideas can be internally transformed by us in various ways, suggesting (for example) new sorts of 'room'. Some transformations will be more radical – and potentially more creative – than others. To transform a frame by changing a low-level slot-filler is less creative than to redefine the frame at a high level. So to change a room by substituting wallpaper for paint is to make a less fundamental change than to add a second door and/or window. To do away with walls and doors altogether is more fundamental still. The idea of open-plan housing was creative, for it challenged some of the most fundamental assumptions in the concepts of 'house' and 'room'.

If Kekulé's mind contained a conceptual structure something like a molecule-frame, then *string* must have been represented as a defining property, not as a slot whose filler could vary. Changing *string* to *ring* would then be broadly comparable to defining a *room* in terms of floor-levels and pathways rather than walls and doors.

An architect can focus on functional, instead of physical, issues. If what one needs to fulfil the various domestic functions carried on in a house is a number of clearly defined and interconnected spaces, then physical walls are not always necessary. Indeed, 'focus on the function' is a heuristic that is very often used by architects, designers and creative engineers to escape from stereotyped thinking.

Since benzene is neither an artefact nor a biological organ, Kekulé could not hope to solve his problem by switching from the physical to the functional. But he might, as suggested in Chapter 4, have applied the heuristic 'consider the negative' to the relevant defining-property of his molecule-frame.

In fact, Kekulé's own report suggests that the visual imagery of tail-biting snakes provided an important clue. If so, his creativity was due in part to *analogical* representation.

An analogical representation is one in which there is some structural similarity between the representation and the thing represented. Moreover, the similarity (whatever it may be) is a significant one. That is, it is specifically exploited in interpreting the representation. (Otherwise, it would be a mere idle similarity: a matter of objective fact but of no psychological interest, like the similarity between the constellation Cassiopea and the letter W.)

To understand an analogical representation is thus to know how to interpret it by matching its structure to the structure of the thing represented in a systematic way. In general, ways of thinking (inference procedures) that are normally associated with one of the two structures are transferred to the other.

Some homely examples of analogical representation are maps, diagrams, scale-models and family-trees (wherein family relationships of blood and marriage are represented by verticality, horizontality, juxtaposition and connectedness). More specialized examples include the chemist's 'periodic table' (which maps abstract patterns of chemical properties on to a spatial lattice), and Longuet-Higgins' harmonic arrays and boxes.

An analogical representation need not use space. The results of a race involving twelve athletes could be represented analogically by space (listing their names in a row or column, or writing them inside circles of increasing area), by time (reciting them in order), by number (the first twelve integers), by the alphabet (A to L), by sound (of heightening pitch), or even by colour (twelve pink/red beads of deepening hue).

But it is no accident that many analogical representations are spatial, for vision is our most powerful sense. The visual system has evolved to notice spatial relations like connectedness, juxtaposition and gaps, and to see connectedness as implying a possible pathway along which the eye and/or body might move in either direction. We remarked in Chapter 4 that the gaps in the periodic table prompted chemists to search for new elements. What we did not say there is that it is *natural* for us to notice spatial gaps. Likewise, it is natural for us to notice spatial similarities and symmetries.

Suppose that a certain conceptual space, or problem-domain, can be represented as a spatial model or diagram. In that case, we can exploit our everyday powers of visual processing to see relationships, gaps and potential pathways which otherwise we might have missed. And if we can do that, we can explore the conceptual space more easily.

No wonder, then, that Kekulé *saw* the similarity between tail-biting snakes and closed strings forming rings (though his 'seeing' the chemical significance involved other types of inference, as outlined in Chapter 4). And no wonder that Longuet-Higgins ignored the octave in defining major and minor keys. For if a key is two-dimensional, it can be represented by a spatial array – with all the inferential power that naturally provides.

Writers on human creativity usually take our highly sophisticated powers of visual processing for granted. But AI-workers trying to model our use of diagrams cannot. A computer's inferential powers do not come 'naturally', and providing computers with the ability to interpret visual images is no simple matter. Consequently, many programs that appear to be using spatial representations either do not use space at all (but numbers, instead), or use rudimentary procedures for inspecting simple spatial arrays in what, by human standards, are very limited ways.

Longuet-Higgins' program, for example, could find the next-door-neighbours (up, down, left or right) of a note. But it was blind to the symmetries, or repeating patterns of notes, that we can see when we look at the array in Figure 5.2. Nor could it (or any program like it) have 'seen', or inferred, from Figure 5.5 that object B, if tipped at the top right-hand corner, would fall on to D, whose lower left-hand corner would come to rest on the ground.

Some programs exist which can see this, and which do so (in effect) by taking successive 'snapshots' of the local changes in the image-array that would ensue if the first object fell through the space separating it from the second. But even these systems cannot see the similarity in shape between B and C (although they could fairly easily be extended so as to do so).

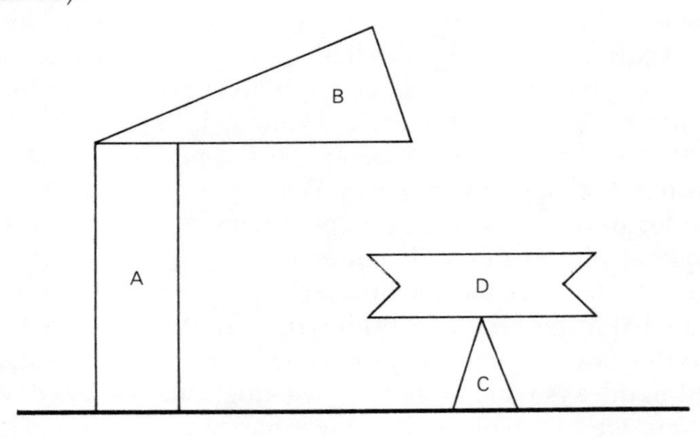

Figure 5.5

As we learn more about the processes that make human vision possible, so we shall be able to include them in programs designed to use spatial diagrams. Meanwhile, our computer models *and our psychological understanding* of those cases of human creativity which involve visual representation will be limited. In short, Kekulé's snakes are more puzzling than they may seem.

Nor do Kekulé's snakes show that imagery is invariably useful. In truth, it is not. Any representation can block creativity, as well as aiding it. To be creative is to escape from the trap laid by certain mental processes currently in use. People can be trapped not only by a frozen heuristic, but also by a frozen representation.

Many trick-puzzles depend on this. For example, if you try the following mathematical puzzle on some friends chatting over coffee, you may find that the first person to solve it knows the least about mathematics. Engineers and physicists, for instance, usually have trouble with it (even though one might expect them to guess that there is some catch). Indeed, two world-famous mathematicians on whom I have tried it each refused repeatedly to answer, insisting that while the principle of solution is obvious the solution cannot in practice be found without either a computer or lengthy pencil-and-paper calculations.

Here it is: *There are two houses, x feet apart. A twenty-foot string is suspended between two points, A and B, on the neighbouring walls of the houses. A and B are at the same height from the ground, and are high enough to allow the string to hang freely. The vertical 'sag' in the string (the distance between the string's lowest point and the horizontal line joining A and B) is ten feet. What is x? That is, how far apart are the houses?*

As with the necklace-game, it may be illuminating if you attempt to solve the puzzle yourself before reading further. (If you are wondering whether you are allowed to draw or write on a scrap of paper, the answer is yes). Then, try it out on some friends.

A mathematician, engineer or physicist is likely to say, 'Aha! It's a catenary curve. Here's the equation. But I can't possibly tell you what x is: I can't work out catenaries in five minutes.' If you assure them that the puzzle can be solved without giving up the whole afternoon, this may not help. They may be so blocked by the idea of catenary curves that they cannot break away from it. If they do, they are likely to try trigonometry (warning you, as they pore over their diagram, that they will be able to provide only an approximate answer). Eventually, they will abandon that approach too.

Less mathematically sophisticated people usually start by drawing the string and the houses, or try to visualize them 'in their mind's eye'. This doesn't work either.

Many people can go no further, declaring the problem to be insoluble. But it isn't. My son solved it instantly. (Do you need a clue? Concentrate on the ten and the twenty. Do you need the answer? I have nothing to say.)

If, by now, you have solved the puzzle yourself, you can see why drawing (or imagining) a diagram can lead one *away* from the solution.

(One friend who solved this problem very quickly without pencil-and-paper – we were climbing Snowdon at the time – described himself as 'a very bad visualizer'.) People often say, sometimes citing Kekulé's writhing snakes as an example, that visual imagery aids creativity. So it may. But it can also prevent it.

The 'blocking' effect of visual representation is evident also from the strange case of Euclid, Pappus and the geometry-program.

The geometry-program was a very early AI-system, designed to demonstrate theorems in elementary Euclidean geometry.[4] Its general strategy was to work backwards from the theorem to be proved, using what AI-workers call *means–end analysis* (or *planning*) to represent the problem on several hierarchical levels of goals and sub-goals.

Taking the required theorem as its main goal, it would first try to find an expression (or a conjunction of expressions) from which the goal-theorem could be immediately inferred. If that expression (or every item within the conjunction) was already listed as an axiom or a previously proved theorem, all well and good. The problem was solved.

If not, the program would set up the sub-goal of finding a lower-level expression (or conjunction) from which the higher-level one could be directly inferred. If that expression did not consist solely of axioms or theorems, then a sub-sub-goal would be set up to prove it in turn . . . and so on. (On average, the program used eight levels in the goal-hierarchy to solve the problems it was given.)

The program enjoyed a degree of geometrical 'insight', without which its problems would have been unmanageable. That is, it used heuristics to decide on the most promising paths in the search-space. There are always very many different expressions from which a given goal-theorem (at whatever level) *could* be immediately inferred. The problem-solver's task is to select one that can be derived, in a reasonably small number of steps, from Euclid's axioms (and any previously proved theorems). Instead of attempting an exhaustive and, in practice, impossible search through *all* the candidate expressions, the program relied on a number of heuristics to focus on the more likely ones.

Its most important heuristic method was to check its incipient suggestions (the candidate expressions) against diagrams representing the geometrical constraints mentioned in the problem. (The initial diagram for a given problem was provided 'free' to the program, as it often is in geometry textbooks.) Any expression that conflicted with the diagram – for instance, mentioning a right-angle where none existed in the diagram – was assumed to be false, and was abandoned.

If all else failed, the geometry-program could call on some heuristics designed to extend the search-space. That is, it could transform the diagram, using a method of geometrical construction which Euclid himself employed.

In essence, it could draw a new line connecting two previously unconnected points in the diagram, and it could extend this line to intersect with any original lines lying 'in its path'. It would give new letter-names to the new intersection-points, as the human geometer does.

For instance, in order to prove that the opposite sides of a parallelogram are equal, it constructed the diagonal – shown as a dotted line in Figure 5.6(a). Again, it considered the line joining the mid-points of the diagonals of a trapezoid (a quadrilateral with two parallel sides): does the extension of this line exactly bisect the 'non-parallel' side of the trapezoid? In proving that it does, the program constructed the dotted line shown in Figure 5.6(b) and named the new intersection-point, K.

Figure 5.6 (a)

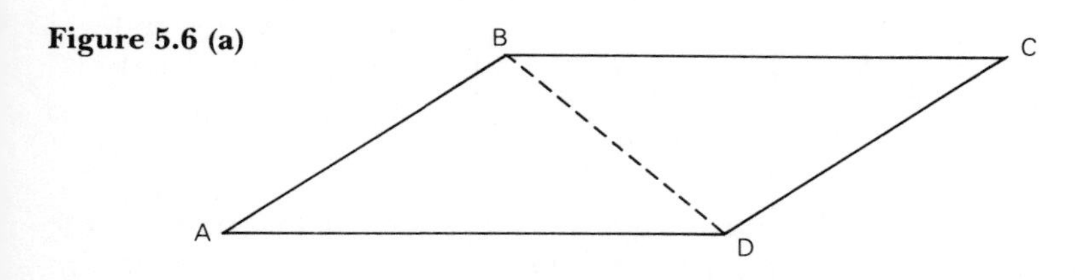

Figure 5.6 (b)

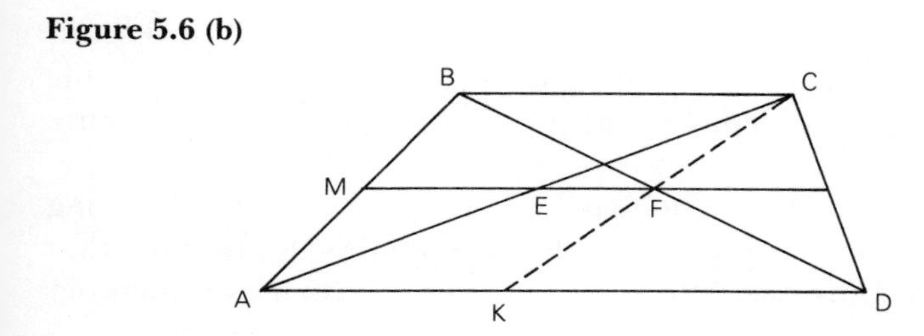

The geometry-program was able to prove many theorems in elementary Euclidean plane geometry. Over fifty had been achieved within less than a year of the first experiments with the system. One is especially interesting here: its proof that the base-angles of an isosceles triangle are equal.

When Euclid wished to prove that the base-angles of an isosceles triangle are equal, he used a fairly complicated method. The diagram he started with is shown in Figure 5.7, and the diagram he constructed – which looks something like a bridge – is shown in Figure 5.8.

To transform the one into the other, he first extended the two sides of the given triangle (so AB and AC were extended to AD and AE). Next, he named an arbitrary point (F) on one extension. Then, he identified a point (G) on the other extension, such that the two points (F and G) were equidistant from the apex of the triangle. Finally, he constructed two extra lines (FC and GB), each of which joined one of the newly named points with the opposite end of the base of the original triangle.

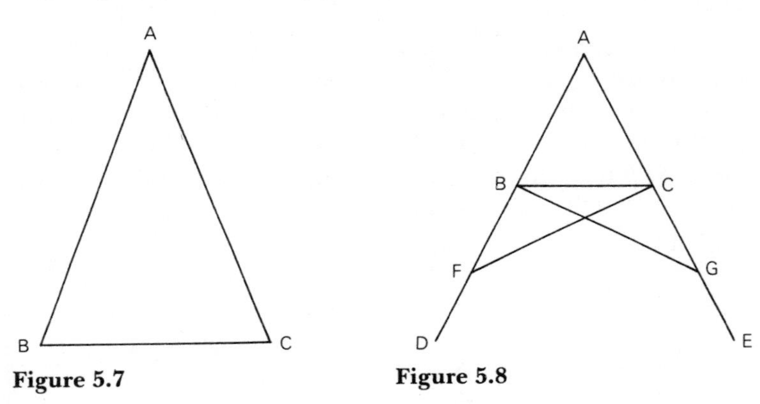

Figure 5.7 **Figure 5.8**

Having done this, Euclid argued at some length to arrive at the theorem he wanted to demonstrate. The details of his proof need not concern us. (It is the first difficult proof in his *Elements of Geometry*, occurring very early in Book I, and is called the *pons asinorum* because it eliminates the 'asses' who cannot 'cross the bridge'.) The important point, for our purposes, is that Euclid introduced considerable complexity into the problem by transforming Figure 5.7 into Figure 5.8.

Because of the difficulty of the *pons asinorum* proof, children learning geometry at school are taught to use a different method for proving that the base-angles of an isosceles triangle are equal. (You may remember it from your own schooldays.) But this method, too, involves construction.

Specifically, the apex-angle is bisected by a line which intersects the base of the triangle, so that Figure 5.7 is converted into Figure 5.9. The proof, which is much simpler than Euclid's, then goes like this:

Consider triangles ABD and ACD.
AB = AC (given).
AD = DA (common).
Angle BAD = angle DAC (by construction).
Therefore the two triangles are congruent (two sides and included angle equal).
Therefore angle ABD = angle ACD.
Q. E. D.

(One cannot accuse Euclid of 'missing' this simple schoolroom-proof. He could not have used it early in Book I, instead of the *pons asinorum*, because he had not proved any theorems about congruence; these occur much later in his geometry.)

Figure 5.9

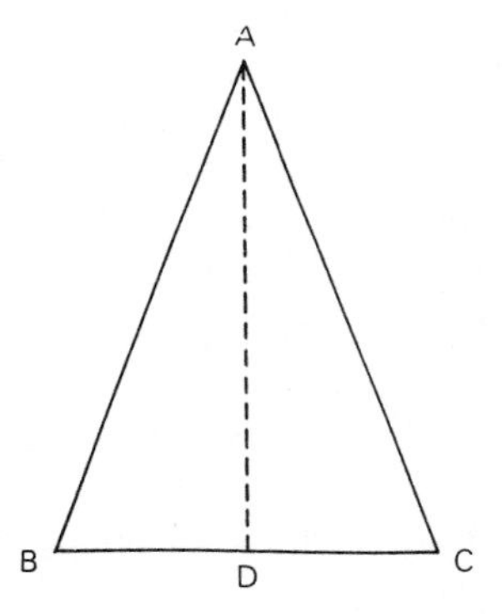

The geometry-theorem prover, which knew about congruence, might have been expected to do the same sort of thing. However, it did not.

The program allowed for construction only if all other attempts to solve the problem had failed. Accordingly, it engaged in a thorough search of the unreconstructed search-space – a search which succeeded. Instead of altering the diagram in any way, it used Figure 5.7 and one of Euclid's theorems concerning congruence to argue, in effect, like this:

Consider triangles ABC and ACB.
Angle BAC = angle CAB (common).
AB = AC (given).
AC = AB (given).
Therefore the two triangles are congruent (two sides and included angle equal).
Therefore angle ABC = angle ACB.
Q. E. D.

This proof is much more elegant than Euclid's: no 'bridge' is needed. And it is more elegant than the school-taught version, which also uses the concept of congruence, because it involves no construction of added lines.

As remarked in Chapter 3, the schoolchild who produced this proof would be regarded as P-creative. What of the computer? Does its performance even appear to be P-creative?

It would be misleading to say (with Lady Lovelace) that the programmer 'ordered' the computer to produce this simple congruence-proof, for he was just as surprised by it as anyone else.

Nor should we simply say 'Creativity? Of course not! It was able to produce the proof only because of its program,' and leave it at that. For this no-nonsense objection, which assumes that genuine creativity could never reside in a program, concerns the last of the four Lovelace-questions distinguished in Chapter 1 – namely, whether any conceivable computer could *really* be creative. Our interest here is in (a special case of) the second Lovelace-question, whether a specific computer *appears* to be creative.

At first sight, it seems that the answer in this case must be 'Yes'. After all, the computer came up with a proof much simpler than Euclid's, and simpler even than the commonly used proof. It seems to satisfy Poincaré's criterion of mathematical insight, that something 'gratifies our natural feeling for mathematical elegance'.

However, we have seen that creativity is a matter of using one's computational resources to explore, and sometimes to break out of, familiar conceptual spaces. On closer inspection, it is clear that the program did not break out of its initial search-space. It did not even bend the rules, never mind break them.

Before considering why this is so, let us look at how Pappus of Alexandria solved the problem, six centuries after Euclid. Pappus also produced a congruence-proof of the 'base-angles equal' theorem without doing any construction. Indeed, his proof – as written down on parchment or paper – was identical to the proof produced by the geometry-program. But, unlike the program, Pappus had to *escape* from seeing something about Figure 5.7 which we can see too (but which the program was not able to see).

On looking at Figure 5.7, what do *you* see? Presumably, a triangle: an identifiable bounded area of a certain shape. Suppose I ask you to consider *two* triangles? Presumably, you will imagine, or draw, two distinct (and probably non-overlapping) shapes: something like Figure 5.10, perhaps. And suppose you were asked – or had it in mind – to use some geometrical theorem (about congruence, for instance) which concerns two triangles? Presumably, you would assume that the theorem is to be applied to a diagram something like Figure 5.10.

Figure 5.10

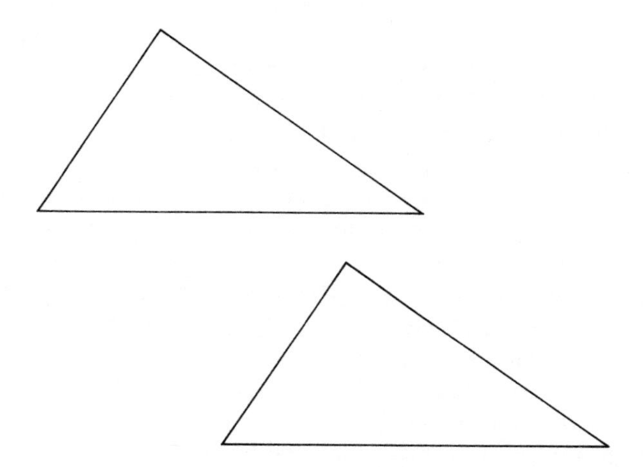

I have as much right to my 'presumablys' here as anyone does to presume that Kekulé could have noticed the similarity between snakes and strings, or between tail-biting snakes and rings. All these presumptions take for granted the normal functioning of the human visual system.

Pappus (like our hypothetical P-creative schoolchild) enjoyed these functions too. But he was able to set them aside. He realized that the concept of congruence could be applied not only to two separately visualized triangles but also to *one and the same triangle, rotated.* (He imagined the triangle lifted up, and replaced in the trace left behind by itself.)

The geometry-program did no such thing – and it *could not* have done it, no matter how long it was allowed to run. The reason lies in its method of representation.

Given our previous discussion of spatial representation, you will not be surprised to hear that this very early AI-system did not use 'real' diagrams. Instead, it represented its triangles abstractly, by numbers identifying points in a coordinate-space (comparable to the x, y coordinates on a graph).

Similarly, it did not identify angles visually, by their vertices and rays, but by an abstract list of three points: the vertex, and a point on each ray. Consequently, it represented what we see as 'one' angle as two (or more) differently named angles. In Figure 5.6(b), for example, it would have identified angles ABD, DBA, MBD, DBM, ABF, FBA, MBF and FBM as distinct angles. (To prevent the program's getting thoroughly bogged down in *proving* that each of these angles is equal to every other, a procedure was eventually added allowing it to *assume* their equality, since it could not *see* it.)

It was because of this abstract, non-visual method of representation that the geometry-program was able to consider the congruence of triangles ABC and ACB in Figure 5.7. For this reason, too, it could not recognize the interest of the proof it produced.

It did not have to escape from the 'two-separate-triangles' constraint normally attached by us to theorems about congruence, because it could not see separate triangles, as we do. It had no analogue of Pappus' (visual) geometric intuition that the mirror-image of an isosceles triangle must be coincident with itself. For its 'geometrical intuition' consisted entirely in abstractly defined heuristics, as opposed to ways of exploring real spatial structures.

Moreover, it could not have had the idea of *rotating* any triangle, since it knew nothing about the third dimension. One might even say that part of Pappus' creativity lay in his using the third dimension to solve a problem in *plane* geometry. (One might also say that Pappus cheated, since lifting-up a triangle and replacing it in the same area is not a plane-geometry procedure. But creatively breaking the rules, or even bending them, could always be called 'cheating'.)

Unlike Pappus, the program did not transform its initial search-space in any way, not even by construction. Rather, it did something which shows that a result we might have thought to lie only in some other conceptual space – one containing Figure 5.8 – could be reached by a thorough exploration of that one.

In this case, then, Lady Lovelace has a point. Clearly, one should not take her words to show that what a computer program can or cannot do is always intuitively obvious. It is not. But despite the geometry-program's capacity to surprise us, it does not have a strong case to be called creative.

Many cases of creativity depend, at least in part, on the mind's associative power. You may feel that the concepts discussed in this chapter are incapable of explaining this. Admittedly, we have considered 'associative' (semantic) nets. However, it is one thing to describe the human memory as a network of meanings. (Who would disagree? Certainly not Coleridge, Poincaré or Koestler.) It is quite another to try to explain memory in terms of semantic nets as traditionally understood in AI.

Again, it is one thing to say that people use heuristics to solve problems. It is another thing to say that all the challenges facing creative thinkers can be met by applying ordered heuristics to strictly defined search-trees, in the way that traditional AI problem-solvers do.

Certainly, some real-life problems involve relatively cut-and-dried search-spaces: the necklace-game, Euclidean geometry, chemistry, harmony, chess. But others apparently do not. Moreover, even those problems which do involve strict rules may also require mental processes of a more flexible kind.

Chess-masters, for instance, have to be able to recognize thousands of different chess-positions. Kekulé not only had to know the rules of chemistry, but also had to be able to see a tail-biting snake as a closed ring. And literary and musical artists (and audiences) often need to be able to recognize some phrases as being reminiscent of others. Many creative acts, as Koestler pointed out, involve 'seeing an analogy where no one saw one before'. How is this possible? In Koestler's words, 'Where does the hidden likeness hide, and how is it found?'

Koestler's question deserves an answer, which rigidly defined search-trees and semantic nets cannot provide. In many real-life problems, a large number of constraints must be met *en masse* but none is individually necessary: each constraint 'inclines without necessitating'. (Think, for instance, of everyday achievements like recognizing a friend, irrespective of changes in hairstyle, suntan, acne and all.) Often, the same is true of the problems facing scientists and (especially) artists. Harmony involves rules, to be sure; and sonnet-form does too. But is it even conceivable that poetic imagery, for example, might be explained in computational terms?

6

Creative Connections

The splendid unicorn depicted in the 'Lady and the Unicorn' tapestries in Paris has the body of a horse, cloven feet like a bull, the bearded head of a goat, a lion's tufted tail and the long horn of a narwhal. Unhappily, given its poignantly sweet expression, no unicorns exist. Never having seen one, how did the embroiderers, or the myth-makers by whom they were inspired, come up with the idea?

And what about the water-snakes described in Coleridge's poem *The Ancient Mariner*: 'blue, glossy green, and velvet black', they 'moved in tracks of shining white' shedding 'hoary flakes' of 'elfish light'? Not only had Coleridge never seen them, he had not seen any other exotic sea-creatures either (he had never been to sea). And although 'water-snakes' had figured occasionally in his voracious reading, he had come across none with 'hoary flakes'.

What he had come across was a wide variety of sources in various languages, ranging from Captain Cook's diaries and many other memoirs of sea-voyages (some published centuries before), through poems published by others or half-written by himself, to technical treatises on optics and the variegated volumes of the *Philosophical Transactions of the Royal Society*. What sort of mental mechanism could possibly produce light-shedding water-snakes from such a weird rag-bag? Pulling a rabbit out of a hat is less mysterious.

Whatever this mental mechanism may be, it might seem at first sight that computational concepts could not possibly explain it. Computer programs have a 'feel' about them that is very unlike what artists or scientists say when they try to describe their own creative moments.

To be sure, people can describe only their conscious thoughts – who knows what unconscious influences may be at work? Even so, the tenor of most introspective reports is at odds with the heuristic methods used by the geometry-theorem prover (and with the exploratory procedures of many 'creative' programs, including most of those to be described in Chapters 7 and 8).

The molecular biologist François Jacob, for example, put it like this:

> Day science employs reasoning that meshes like gears. . . . One admires its majestic arrangement as that of a da Vinci painting or a Bach fugue. One walks about it as in a French formal garden. . . . Night science, on the other hand, wanders blindly. It hesitates, stumbles, falls back, sweats, wakes with a start. Doubting everything. . . . It is a workshop of the possible . . . where thought proceeds along sensuous paths, tortuous streets, most often blind alleys.[1]

There are many ideas which this description of creative science brings to mind, but 'program' and 'computer' are most unlikely to be among them.

You may feel that, from the biological point of view, this is hardly surprising. Why should anyone think that 'night science', or the poetic imagination, or the depiction of mythical beasts, involve psychological processes that can be captured by programmed rules? Brains are very different from computers. So it is only to be expected that people can do things which computers cannot do.

Probably (you may allow), computer-buffs could specify rules to model the routine problem-solving of 'day science'; indeed, some successful programs for routine scientific use already exist. Very likely, they could program a quasi-mechanical search through a dictionary for matching metre and rhyme: 'water-snakes' and 'hoary flakes'. Possibly, to give them the benefit of the doubt, they could model certain sorts of scientific originality: inductive methods might discover hidden patterns in data, and might even come up with some simple mathematical laws. And certainly, any AI-student could write a concept-shuffling program to churn out a family of non-existent monsters: input *horse*, *goat*, *bull*, *lion*, *narwhal* and also *body*, *head*, *feet*, *tail*, *horn*, crank the handle – and we could get a unicorn. But we would get the beast without the myth (and without the sweet expression). Add *fish* and *woman* and we might get mermaids; but we would not hear them singing.

In short (so the objection goes), it is folly to expect scientific and poetic creativity to be explained in computational terms. 'Computation' means 'following a program', and whatever the brain is doing it is surely not that.

The key insight in this objection is that the brain is very different from the digital computer. The designer of the digital computer, John von Neumann, was well aware of this, and insisted that 'the logic of the brain' could not be like computer programs. (He suggested that it might be something like thermodynamics – an idea that is causing much excitement today, as we shall see.) Any scientific account of creativity that does not recognize the difference between brains and digital computers is doomed to failure.

The key point in the reply is that brains are, to some extent, *like* a certain type of computer-model: namely, connectionist systems or neural networks. 'Computation' in connectionist systems does not mean following a program in the traditional sense. We shall see that the ideas used to describe connectionist computation are helpful in understanding how the brain works – and how some aspects of human creativity are possible.

However, first things first. Consider those water-snakes. They cavort not only in the poem but also in a masterly literary detective story tracing the sources of Coleridge's imagery.[2]

Most of the clues in this literary thriller, written over sixty years ago by John Livingston Lowes, were found in Coleridge's own notebook, in which (over a period of three years) he had jotted down sundry passages and ideas drawn from his exceptionally wide reading. The scholar-sleuth goes to the original texts (*Purchas's Pilgrimage* for instance, or the *Philosophical Transactions*) to examine the initial context of the scribbled phrases, and even rummages in the volumes mentioned in the footnotes there. Often, he finds distinct intellectual footsteps, persuasive indications that the poet had wandered along the very same path.

Livingston Lowes provides detailed evidence for specific conceptual associations in Coleridge's mind which probably underlay specific words, lines or stanzas.

For reasons we shall discuss in Chapter 9, 'evidence' and 'probably' are the best we can expect in investigations of this kind. Usually, the scholar must present his case to the civil, not the criminal, courts: the origin of a particular line or image can sometimes be established beyond all reasonable doubt, but more often we have to do with the balance of probabilities. Certainty is available only when we have a confession from the poet – such as Coleridge's identifying a sentence from *Purchas's Pilgrimage* as the source of the initial frame and some of the imagery in *Kubla Khan*.

For our purposes, this does not matter. We are like the crime-prevention officer (who asks how burglaries are possible) rather than the detective (who asks whether Fred Bloggs stole the diamonds). Admittedly, the crime-prevention officer must be able to produce some plausible stories about how individual burglaries might have been committed. But he does not need absolute proof. Likewise, we need to show only that a specific idea probably arose, or could have arisen, in a particular way, since our question is how it is possible for creativity to occur at all.

The ancestry of the water-snakes is one of the mysteries clarified by this painstaking literary scholarship. Livingston Lowes locates their origins in at least seven volumes traced by specific passages from the poet's notebook, and in other books, too, which Coleridge is known to have read. He offers many pages of careful detail and subtle argument (and pages of fascinating footnotes) to display their genealogy as completely as he can. Let us note just a few points here.

'Snakes' and other sea-creatures of 'Greene, Yellow, Blacke, White . . .', figure in a late-sixteenth-century description of the South Sea (quoted by Purchas in 1617), and 'water-snakes' appear in Captain Dampier's journal almost exactly a hundred years later. A hundred years later still, in a long and turgid poem called *The Shipwreck*, a contemporary of Coleridge referred to the 'tracks' of 'sportive' dolphins, and to porpoises who 'gambol on the tide' and whose 'tracks awhile the hoary waves retain'.

A book about Lapland, printed as juxtaposed columns of Latin and Norwegian, describes dolphins with the phrase 'in mari ludens' (playing in the sea). Moreover, a *Philosophical Transactions* paper on 'Luminous Appearances in the Wakes of Ships' reports 'many Fishes playing in the Sea' making 'a kind of artificial Fire in the Water', and mentions 'fishes in swimming' who 'leave behind 'em a luminous Track'.

The chapter on 'Light from Putrescent Substances' in Priestley's *Opticks* records fishes which 'in swimming, left so luminous a track behind them, that both their size and species might be distinguished by it'. And finally (for us, though by no means for Livingston Lowes), Captain Cook mentioned 'sea snakes', and saw sea-animals 'swimming about' with 'a white, or shining appearance', who in the candlelight looked 'green' tinged with 'a burnished gloss' and in the dark were like 'glowing fire'.

In case these fragments ring no bells in your memory, here again is the description of the water-snakes which appeared earlier in this chapter: 'blue, glossy green, and velvet black', they 'moved in tracks of shining white', shedding 'hoary flakes' of 'elfish light'.

Livingstone Lowes does not suppose that Coleridge's deliberate search for references to sea-creatures was, in itself, enough to produce the imagery of *The Ancient Mariner*. Indeed, he criticizes an earlier, and unsuccessful, poem in which Coleridge used these bookish references with hardly any transformation. (In general, what Hadamard called the preparation phase involves deliberate search; inspiration comes later.)

But the preparatory search established a field of meaning, from which (as he explains) the poetic description of the water-snakes was created by the poet's 'extraordinary memory' and 'uncanny power of association'. 'Uncanny' here does not mean alien, for Livingston Lowes saw Coleridge's mental powers as better developed than other people's, not radically different (a view defended in Chapter 10).

Coleridge's poetic imagination, he says, structured individual stanzas and the poem as a whole. But the genesis of the water-snakes was associative, a spontaneous – and only partly conscious – result of what Coleridge himself called the 'hooks and eyes' of memory. Thus a single telling word ('hoary') was remembered from a mass of turgid verse – but other words in the context (such as 'tracks', 'supportive', 'gambol' and even 'dolphin') facilitated the relevant associations. In sum, Livingston Lowes described the poet's mind (and other minds too) as a richly diverse and subtly associative conceptual system.

You may feel that the water-snakes are less surprising, less creative, than the unicorns. After all, water-snakes exist, and had been described before Coleridge wrote about them. But there is an interesting feature of the water-snakes (and, on close examination, of unicorns too) which makes their explanation more tricky than it may appear.

Up to a point, one might describe the creation of unicorns as a matter of conceptual cut-and-paste. Indeed, some of Livingston Lowes' own explanatory language (about breaking down, separating and re-combining ideas) has this flavour about it. Insofar as that is what creativity involves, a traditional program of a rather boring kind might mimic it with some success. The computerized monster-generator mentioned above, for example, could give us unicorns, mermaids, centaurs and many more. Agreed, there is also evaluation, in selecting the unicorn rather than some other conjectured beast; and there is a rich penumbra of myth and magic surrounding unicorns. Cut-and-paste can explain neither the evaluation nor the myth (nor the neck, as we shall see shortly). But *coming up with the novel idea* may seem relatively unproblematic.

The water-snakes are less amenable to cut-and-paste explanations. Admittedly, the poet uses expressions 'cut' from different sources:

'water-snakes' and 'hoary', for instance. But the sources are so different, and so diffuse, that the spontaneous association of these ideas is a phenomenon needing explanation in itself. How does the mind locate the particular ideas involved?

Moreover, most of the associated ideas are not pasted together so much as merged and subtly transformed. Coleridge's 'glossy green' does not appear as such in any source, but 'green' or 'Greene' occurs in several, and 'gloss' occurs close by in one; likewise, the 'luminous Track' and 'shining appearance' were initially reported of fishes, but in the poem it is the water-snakes who 'moved in tracks of shining white'. The ideas aroused by the original sources are not used as bits in some conceptual mosaic, but are blended to form a new image. (Indeed, the same applies to unicorns; the goat's head and the horse's body are joined in the tapestries by a neck typical of neither goat nor horse, but only of unicorns.)

In such cases of creative merging or transformation, two concepts or complex mental structures are somehow overlaid to produce a new structure, with its own new unity, but showing the influence of both. How can this be?

Livingston Lowes was well aware that cut-and-paste cannot suffice to explain this sort of creative novelty. As he put it: 'the creatures of the calm are not fishes + snakes + animalculae, as the chimaera was lion + dragon + goat. No mere combination of entities themselves unchanged explains the facts.' Conscious recollection and reconstruction have an important place in creativity, but they are not enough.

He continues: 'The strange blendings and fusings which have taken place all point towards one conclusion, and that conclusion involves operations which are still obscure.' The origin of creativity is the unconscious mind – not the Freudian unconscious of repressed instinct, but what Coleridge himself called 'that state of nascent existence in the twilight of imagination and just on the vestibule of consciousness'. We have to do with the subtle, unconscious processes of the imagination. Again, he quotes Coleridge: 'The imagination . . . the true inward creatrix, instantly out of the chaos of elements or shattered fragments of memory, puts together some form to fit it.' The root of poetry (and, he says, of science too) is *unconscious association*, a process which can re-form ideas as it associates them.

To identify unconscious association as a creative principle is one thing, and to say how it works is another. Livingston Lowes acknowledged that the 'operations' of the unconscious are 'still obscure'. He is convincing when he specifies the diverse ideas associated in Coleridge's memory, and when he compares these source-ideas with their newly formed descendants. He is convincing, too, when he rejects mere recombination as a 'crassly mechanical explanation'. But he can offer only intuitive and metaphorical accounts of how the memory functions.

He concentrates on the raw material and the poetic results of unconscious association, not on its fundamental mechanism. He insists that creativity is universal and non-magical, that it is a natural feature of the human mind that can be understood in psychological terms. But his interest is not the same as ours, which is to find some way in which creativity might be scientifically understood. In short: just what are the hooks and eyes of memory, how do they find each other, and how can they fit together to produce a novel form?

Many people today, asked whether computer science could help answer these questions, might borrow a scornful phrase from Livingston Lowes in dismissing such 'crassly mechanical explanation'. In justification, they might appeal to biology. Brains, they may insist, hold the secrets of poetry. Computers, being unlike brains, are irrelevant.

Those who favour this objection usually regard it as *obvious* that the brain (with its millions of richly interconnected neurones) can support associative thinking. Perhaps so. But a deep chasm divides 'can' from 'how'. Just how the brain supports association is not obvious at all. It is not obvious, for instance, just how the sorts of creative association and merging described by the literary critic can come about. Could computational concepts help us understand how poetry is possible?

To answer this question we must consider connectionist computer-models: how they work, and what they can do. Connectionist systems are used for psychological (and neuroscientific) research, and in technology too – for automatic face-recognition, for example. They are parallel-processing systems whose computational properties are broadly – *very* broadly – modelled on the brain. They are not programmed so much as trained, learning from experience by means of 'self-organization'. And, as we shall see, they can do some things which are crucial to creativity.

In asking whether connectionist ideas help us to see how human creativity is possible (the first Lovelace-question distinguished in Chapter 1), what these systems could do *in principle* is what counts. Even so, what they already do in practice is intriguing, and it strongly suggests (in reply to the second Lovelace-question) that computers might indeed do things which appear to be creative. It suggests also (in response to the third Lovelace-question) that computers might recognize certain aspects of creativity, being able – as Turing suggested – to prefer 'a summer's day' to its wintry alternative.

One intriguing feature of connectionist systems is their capacity – in principle, and to some degree already in practice – for 'pattern-matching'. They can recognize a pattern that they have experienced before (much as you can recognize a face, an apple or a postage-stamp). Moreover, their pattern-matching is highly flexible in various ways, which find ready analogies in human minds but not in traditional computer programs.

For example, connectionist systems can do 'pattern-completion': if the current input-pattern is only a part of the original one, they can recognize it as an example of *that* pattern. (Likewise, you can identify an apple with a bite out of it as an apple, or a torn stamp as a stamp.) They show 'graceful degradation' in the presence of 'noise': if a pattern is input again in a slightly different form, they can still recognize it as an example of the original pattern. (Compare seeing a Cox's Orange Pippin after a Granny Smith, or a stamp overprinted with a postmark.)

They can do 'analogical pattern-matching'. That is, an input pattern can call up a range of stored different-yet-similar patterns, whose activation-strength varies according to their similarity (as apples are strongly reminiscent of oranges and pears, and to a lesser degree of bananas).

In addition, these systems have 'contextual memory': an input pattern can activate not only a similar pattern, but also some aspects of its previous context. This is especially true if those aspects have already been partially aroused by the current context. (Similarly, an apple in a religious painting may remind you of Eve, whereas an apple in a still-life does not.)

Another intriguing feature of connectionist models is that they do not need perfect information, but can make do with probabilities – and fairly messy probabilities, at that. In other words, they can compute by using 'weak constraints'. They will find the *best* match to a pattern by weighing up many different factors, none of which is individually essential and some of which are mutually inconsistent. (A person does the same sort of thing in judging the aptness of a poetic image.)

Moreover, many can learn to do better, with repeated experience of the patterns concerned (much as someone brought up in an orchard is more likely to remember apples than someone who has seen apples only once). They learn, and can reactivate, many semantic and contextual associations between different representations. In short, connectionist systems have 'associative memory', grounded in both meaning and context.

Most intriguing of all, they do these things 'naturally', without being specifically programmed to do them. (Likewise, you need not be told about apples; nor do you need explicit rules stating the relation of apples to pears, or even to Eve – although an art-historian may help by telling you that an apple sometimes symbolizes Eve.) Rather, their associative memory and tantalizing, human-like, capacities are inevitable results of their basic design.

The design of one common type of connectionist system can be likened to a class of school-children, asked whether something on the teacher's desk is an apple. The task, in other words, is to recognize an apple as an apple – even though it differs slightly from every apple seen in the past.

These imaginary children are attentive, and each is (just) intelligent enough to know whether her current opinion is consistent with her neighbour's. But they are horrendously ignorant. There is no child who knows what an apple is, and no child who knows the difference between an apple-stalk and an apple-leaf. Instead, each child can understand only one thing: perhaps a particular shade of green (or red, or purple); or circular curves (or straight lines, or sharp points); or matt (or shiny) surfaces; or sweet (or bitter) odours.

Each child chatters non-stop about the tiny detail which is her all-consuming interest. A child's opinion can be directly reinforced or inhibited by the messages she receives from her immediate neighbours – who are talking to children in other desks, who in turn are talking to classmates in more distant parts of the room. So her opinion can be indirectly affected by every child who has something relevant to say. Each child repeatedly modifies her opinion in the light of what her neighbours say (the desks are arranged so that children holding opinions on closely relevant topics are seated near to each other). The more confident she is, the louder she shouts – and the louder her neighbour's voice, the more notice she takes of her.

Eventually, the children's opinions will be as consistent as possible (though there may still be some contradictory, low-confidence details). At that point, the set of opinions considered as a whole is as stable as it is going to get: the classroom is in equilibrium.

The final decision is not made by any one child, for there is no class-captain, sitting at a special desk and solemnly pronouncing 'apple'. It is made by the entire collectivity, being embodied as the overall pattern of mutually consistent mini-opinions held (with high confidence) within the classroom at equilibrium. The stabilized pattern of mini-opinions is broadly similar – though not identical – whenever the class is confronted with an apple (of whatever kind). So the teacher, who knows how the class as a whole behaved when they first saw an apple, can interpret their collective answer now.

This classroom is, in effect, a 'PDP' system (it does Parallel Distributed Processing).[3] The class-decision is due to the *parallel processing* (all the children chatter simultaneously) of *localized computations* (each child speaks to, and is directly influenced by, only her immediate neighbours), and is *distributed* across the whole community (as an internally consistent set of mini-decisions made by all the children).

In PDP models, concepts are represented as activity-patterns across a group of units (children). A given unit can be active in representing different concepts (*apple* and *banana*, perhaps). And a given concept in different contexts (an apple in a still-life, or in a Nativity-scene) is represented by different groups of active units. A PDP-unit does not carry a familiar 'meaning', a concept or idea that can be identified by a single word or easily brought into consciousness. (This is why PDP-processing is sometimes called sub-symbolic.) Rather, it represents some detailed micro-feature, describable only in complex and/or technical language, such as *very pale green at such-and-such a position in the right eye's visual field.*

(Some brain-cells may code information that is more easily expressed, for instance *person walking towards me*. 'Localist' connectionist computer models, such as one to be described in Chapter 7, typically have units coding for familiar concepts, but 'distributed' systems usually do not.)

Slightly different sorts of children correspond to different types of connectionist model – and different computational possibilities. For instance, the children may say only 'yes' or 'no', or they may be able to distinguish 'probably' and 'possibly' as well. And they may always be guided strictly by the evidence available, or they may sometimes speak at random. (This last arrangement is not so silly as it sounds: much as the theory of thermodynamics assigns an infinitesimal probability to a snowball's existing in Hell, so a class of children who sometimes speak at random is *in principle* guaranteed to reach the right decision – although this may require infinite time.)[4] These computational distinctions are relevant to what goes on in the brain. For example, a neurone sometimes fires at random, or spontaneously, without being triggered by its input-neurones.

In many connectionist classrooms, there is a clear distinction between children who can see or smell the thing on the teacher's desk, children who can announce the class's decision as to what it is, and children who can do neither. (In the jargon, these groups are called the 'input' units, the 'output' units, and the 'hidden' units, respectively.)

For instance, one child may be able only to *recognize the presence* of a certain shade of green, while another can only *announce that* the thing is partly green. Indeed, there may be a special row of children (imagine them seated by the left-hand wall) each of whom can perceive some tiny aspect relating to the teacher's question, and another row (by the right-hand wall) each of whom can deliver a tiny part of the class's answer. The job of the children in the middle rows, then, is to mediate between the two wall-rows. Because they communicate only with other children, the children in the middle are in effect hidden from the outside world. The teacher need know nothing about them, because she shows the apple directly only to the left-wall children and listens only to the right-wall children.

Such classrooms can learn from experience, so that they come to associate two different patterns (such as the visual appearance of an apple and the word 'apple'). In the process, the children continually revise the importance they give to the remarks of particular neighbours. A child may decide to pay little attention to a certain neighbour even when she shouts, but to listen carefully to another even when she speaks quietly. These revisions are based on experience. The more often Mary activates Jane, the less energetic she has to be to make Jane take notice of her; and the more often Mary and Jane speak simultaneously, the more likely that Mary will speak up if Jane is speaking.

When the myriad measures of trust accorded by one child to another have stabilized, the internal consistency of the entire set of mini-decisions will have been maximized. In future trials, a maximally consistent class-decision will be reached more quickly, because the relevant pattern of connection-strengths has already been learnt. (Maximal consistency does not mean perfect consistency: there may be some contrary views still expressed within the classroom.)

For example, suppose the class has to learn the full name of the apple on the teacher's desk (which happens to be a Cox's Orange Pippin), so that it can deliver 'Orange Pippin' when the teacher says 'Cox's'. First, comes the lesson. Then, the exam.

In the lesson, the activity-levels of the children in the left-wall row are 'clamped' to represent their hearing the input 'Cox's'. The children in the right-wall now are simultaneously forced to deliver mini-outputs corresponding to the words 'Orange Pippin'. The middle-children are left to chatter to their wallside-neighbours, and to revise their assessments of trustworthiness, until their pattern of activity stabilizes.

In the exam, the left-wall children are clamped (to 'Cox's'), as before. But the right-wall children are not: their activities are now determined not by the teacher but by their neighbours in the middle. When the overall pattern of activity has reached equilibrium, the activity of the right-wall children is taken as the 'answer'. But what counts as equilibrium when the input is 'Cox's' has already been established, during the lesson. Since the levels of trust were in equilibrium (during training) when the right-wall children were being *forced* (by the teacher) to say 'Orange Pippin', equilibrium in the exam-situation is reached only when they are *advised* (by their neighbours) to say 'Orange Pippin'.

(This behaviour is a form of pattern-completion. Similarly, a connectionist system can recognize a torn postage-stamp as a stamp because the entire activity-pattern, originally equilibrated when viewing a whole stamp, is recreated by means of the stored connection-weights.)

If the teacher has a cold when she gives the exam, her voice will sound a little different. Since the input 'Cox's' will not be exactly the same as it was previously, a slightly different set of left-wall children will be strongly activated. But the class as a whole settles down into much the same equilibrium-state as before (it looks for the nearest match, not for a perfect match). Consequently, it finds 'Orange Pippin' as it should. In general, these classrooms respond to family-resemblances between inputs, being able to ignore the slight differences between individual family-members that are perceived.

The very same class of children could learn a different association on the following day: 'Golden' with 'Delicious', perhaps. The reason is that a largely different set of left-wall children will be active when the input is 'Golden', and a largely different set of right-wall children will pronounce 'Delicious'. The overall patterns of mini-opinions will be distinct, and will interfere with each other very little (the children hearing the vowel-sounds for 'o' will be activated in both cases, but the other left-wall children will not).

After this training, then, the class can give the correct response both to 'Cox's' and to 'Golden'. And a third pair of associations could be taught to it next week. Eventually, the class will become 'saturated', as the new patterns interfere with the old. But the time at which this happens will depend on the size of the classroom. The larger the class (the more mini-discriminations it can make), the more pattern-associations it will be able to learn.

In more abstract terms: a connectionist network is a parallel-processing system made of many simple computational units, linked (as brain-cells are) by excitatory or inhibitory connections.

One unit modifies another's activity to different degrees, depending on the relevant connection-weight (expressed as a number between plus-one and minus-one). The details of these weight-changes are governed by differential equations, like those used in physics. A concept is represented as a stable activity-pattern across the entire system.

In networks that can learn, the connection-weights are continually adjusted to maximize the probability of reaching equilibrium. Connections used often are strengthened, and if two units are activated simultaneously then connection-weights are adjusted to make this more likely in future. Specialized input-units and/or output-units can be 'clamped', and the weights in the pool of hidden units are mutually adapted until a maximally stable state is reached. At equilibrium, the highly active units represent micro-features which are mutually supportive, or at least consistent.

These networks are not given, nor do they construct, precise definitions of every concept (pattern) they learn. They gradually build up representations of the broadly shared features of the concept concerned, and can recognize individual instances of the concept despite differences of detail. Moreover, one and the same network can learn several patterns. The larger the network, and the more distinct the patterns, the more associations can be learnt.

People who claim that computational ideas are irrelevant to creativity *because brains are not programmed* must face the fact that connectionist computation is not the manipulation of formal symbols by programmed rules. It is a self-organizing process of equilibration, governed by differential equations (which deal with statistical probabilities) and comparable to energy-exchange in physics. It is the Boltzmann equations of thermodynamics, for example, which prove – what was mentioned above – that a network whose units sometimes fire at random is *in principle* guaranteed to learn (eventually) any representation whatever, much as *in principle* (according to thermodynamics) there could be a snowball in Hell.

In relating creativity to connectionist ideas, we need not delve into the mathematical details of thermodynamics. But we must consider the ability of neural networks to learn to associate patterns without being explicitly programmed in respect of *those* patterns.

Some scientific-discovery programs (as we shall see in Chapter 8) are specifically primed to seek out certain sorts of regularity. A computational system that could pick up a regularity, perhaps a very subtle one, without such pre-knowledge might be closer to actual scientific discovery. Likewise, a mechanism that could spontaneously link – or even merge – concepts from different sources, *via* analogical resonances of various kinds, might cast light on both poetic and scientific creativity.

As an example of non-programmed learning, consider something which all normal children manage to do: forming the past tense in their native language. Infants born into an English-speaking community, for instance, have to learn present/past pairs such as *go/went*, *want/wanted*, *wait/waited*, *hitch/hitched* and *am/was*.

They do this without book-learning, and without consciously thinking about it. Although a few verbs are irregular, most form the past tense in accordance with a grammatical/phonetic rule, such as adding *-ed* to the root-form. But children do not end up with a conscious awareness of the relevant rules, and parents (unless they happen to be professional linguists) could not express the rules even if they tried.

If you listen carefully to what an infant actually says while learning to speak, you will discover a curious thing. At first, the child uses no past tenses at all. When she does start using a few, she uses all of them correctly. Then, as she learns more past tenses, she starts making mistakes which she did not make before. For example, she starts saying *goed* instead of *went*. That is, she over-generalizes: she starts treating irregular verbs as if they were regular ones. Only later, as she learns still more new words, does she revert to the correct form of the irregular verbs which she had used in the first place. By this time, she can also produce plausible past tenses for non-existent verbs: *glitch-glitched*, for instance.

Various subtler regularities are also involved. For example, at one point in the learning process the over-generalization is applied to past-tense words as though they were the relevant present-tense words: *goed* is common early in the error-prone stage, but *wented* is more common later. In addition, some sorts of phonetic change are learnt more quickly than others.

Notice that the child who says *goed*, *wented* or *glitched* is doing something she could not have done before. Initially, she used no past tenses at all, and when she did she produced only those which she had actually heard. Indeed, some linguists have argued that (since the child has never heard these non-words) her behaviour must be due to some unconscious linguistic rule – and we have seen that adding a rule to a system may enable it to generate things it simply could not have produced before.

Moreover, the child picks up various subtle phonetic regularities, which affect her own behaviour without her being consciously aware of them. Much of the expertise that is crucial to creativity likewise involves unconscious knowledge of domain-regularities (in metre and harmony, for instance) picked up in this way.

'Picked up in *what* way?', you may ask. To describe what the child does is not to say how she does it – nor even *how it could possibly be done* by any system.

This is where connectionist modelling comes in. A PDP-network (doing parallel distributed processing) has learnt how to form the past tense of English verbs.[5] On the basis of a number of training-sessions in which it was given present/past pairs, such as *go/went* and *love/loved*, the network learnt to give the past-form when it was presented with the present-form alone. It did this without being told beforehand what regularities to look for, and without formulating them as explicit rules in the process. Moreover, it showed some of the same error-patterns as infants do.

A model such as this shows us *how* an associative system, broadly comparable to the brain, can learn a range of subtleties without being specifically programmed to do so.

The past-tense learner closely resembles the class which learnt to say 'Orange Pippin' when the teacher said 'Cox's'. Each child seated by the left-hand wall can recognize one speech-sound in a particular position (a long 'o' at the end of the word, perhaps, or a hard 'g' at the beginning). Likewise, each child sitting by the right-hand wall can produce one speech-sound in a particular position. A pool of hidden children mediates between the two wall-rows.

During training, verb-forms such as *go/went* are simultaneously presented to the classroom: the relevant left-wall and right-wall children are forced to be very active, and the hidden children are left to equilibrate within these wall-constraints. In the test-phase, activating the left-wall children for the sounds in *go* will (when equilibrium is reached) activate the right-wall children producing the sounds in *went*. Similar processing occurs for *want/wanted*, *wait/waited*, *hitch/hitched*, *am/was*.

Much as our imaginary class could have learnt several more apple-names besides Cox's Orange Pippin, so this actual network learns many different patterns (many different verb-form pairs). Even though the very same set of units learns every verb-pair, the result is not a chaotic babble. The overall pattern of activity at equilibrium is different for each pair.

How do the regular verb-endings get established? The network continually revises the connection-weights of simultaneously activated units, so as to increase the probability of their being activated together. Many verbs – including those ending with the sound-sequence *-ait* – form the past tense by adding *-ed*. In the long run, therefore, a test-input ending in *-ait* will automatically lead to an output ending in *-aited*, irrespective of the initial sound in the verb. Even if it has never heard the verb 'gate', the network will (correctly) produce 'gated'; and if it is given the non-existent verb 'zate' it will (plausibly) produce 'zated'. The same applies when the spoken verb adds *-d* (the written 'e' being silent): having heard 'hitch/hitched', 'pitch/pitched' and 'ditch/ditched', the network will deliver 'glitch/glitched'.

This does not happen by magic, nor because of some supposedly 'obvious' (but actually unspecified) associative process glibly assumed to be carried out by the brain. Nor does it come about by the programming or learning of explicit phonetic rules (such as would be included in a traditional computer program for forming past tenses). Rather, it is due to the gradual revision of the relevant connection-weights, according to the statistical probabilities of certain sound-pairs occurring in the training-input.

The network is not walking in the French formal garden of 'day science', for it is not primed to look out for *-ed*, nor even for changes in word-endings. In principle, it could spontaneously learn any regularities in sets of data-pairs composed of English speech-sounds.

What about the strange pendulum-swing in the network's performance, from correct usage to over-generalization and back again? Given a certain fact about its input, this swing is not strange at all, but an inevitable consequence of the processing principles involved. It is no more surprising than the familiar observation that public-opinion polls are more reliable as the representative sample being interviewed is enlarged.

The verbs used most frequently in the early stages of network-training (and, according to the designers, those used most often by adults when talking to very young children) are irregular: *to be*, *to have*, *to go*, *to see*, *to get* and the like. Only a few are regular (like *wait*). By contrast, the words that are input later are overwhelmingly regular (*hitch*, *pitch*, *restitch*. . . ; *gate*, *debate*, *communicate* . . .).

The earliest word-pairs are learnt as distinct connection-patterns which hardly overlap, and which therefore cannot interfere with each other. As far as the network is concerned, *wait/waited* is no more regular, no more to be expected, than *go/went*. With more experience, more regular verbs are heard. Regular past-tense endings are reinforced accordingly, so much so that they are added to imaginary and irregular verbs (*zated*, *goed*) as well as to regular ones.

If the irregular verbs were not so common, things would end here. But because of the constant repetition of *go/went* (try going – oops! – just one day without using it), the specific connections between these two verb-forms become sufficiently strong to withstand the generalized competition from the regular endings.

Similar remarks could be made about the more detailed speech-errors and learning-patterns in the network's behaviour (such as the blending of *go* with *went*, which leads to the over-generalized *wented*). These errors reflect regularities of a subtle kind, many of which can be described only with the technical vocabulary of phonetics. A few have been observed only in the computer model, and it is not yet known whether those learning-patterns, too, characterize infant speech. If they do, they have eluded the avid ears of psycho-linguists for many years.

A word of caution must be inserted here. One cannot assume that human infants learn in precisely the same way as this model does. Since brain-cells are very much more complex than connectionist units, they presumably do not. Moreover, some critics have argued that children's language-learning has features which network-models cannot achieve.[6] Some of these features concern highly specialized linguistic matters, but others are relevant to many domains. For instance, hierarchical structure is found in grammatical speech and many other sorts of behaviour too, but it is not clear that a connectionist system of this general type could produce such structures. Again, the developmental theory outlined at the end of Chapter 4 implies that children's increasing competence involves spontaneous internal redescriptions of skills they already possess. If this is correct, then statistical regularities in the input cannot suffice to explain how children learn to speak.

For our purposes, these theoretical disputes can be ignored (with one important exception, mentioned below). The main point is that PDP-networks can do things usually assumed to be beyond the capacity of any computer-model. In short, there is more to computational psychology than most people think.

How does all this relate to creativity? Where is the originality, and where are the water-snakes?

Originality is doing something which has not, perhaps even could not, have been done before. In Chapter 5, originality was related to the notion of a search-space, a realm of computational possibilities generated by a particular set of rules. Change the rules, change the search-space – and change the possibilities.

Connectionist systems do not use explicit rules: there is no rule stored anywhere in the verb-learning network stating that verbs ending in *-ait* form the past tense by adding *-ed*. There are, however, clearly defined processing principles which govern the network's computation. And there are specific connection-strengths, *implicitly* coding the probabilities in the data. (These could even be called rules, of a sort; but the word 'rule' is commonly restricted to mean *explicit* representations or instructions, 'effective procedures' to be followed step-by-step – in humans, often consciously.)

Given the absence of explicit rules, one can almost imagine the network saying, with Picasso, 'Je ne cherche pas: je trouve!' But the connectionist *trouvaille* is decidedly less mysterious.

The network carries out no deliberate search (for the past tense of 'go', for example). But the space of possibilities does change, as the pendulum-swing in verb-usage shows. When the network first outputs 'glitched', or even 'waited', it is doing something it could not have done before. As another input-regularity is picked up by the shifting connection-weights, the overall behaviour changes almost as much – though not so suddenly – as when a new axiom is added to a logical system, or an extra search-heuristic to a problem-solving program. (Only 'almost' as much, because the presence of competing connections in the network can in some circumstances suppress the new 'rule': as when *goed* and *wented* both disappear and *went* reappears in their place.)

Moreover, the network picks up dimensions of possibility of a very subtle kind, some of which had not been recognized by anyone before the twentieth century. There is no crude inductive 'priming', no telling the system what to look for before it can find it. Granted, the input-units have to 'hear' the sounds; but scientists have to be able to see and hear too. Since all the connection-weights are initially set at zero, the network knows what phonological features to look for but has no expectation of any specific rule.

In short, this simple network helps us to understand how a richly associative system like the brain could function as 'a workshop of the possible'.

As for the water-snakes, their poetic origins can be better understood by reference to the computational properties of connectionist systems. 'Better understood', that is, in scientific terms.

The intuitive understanding offered by Livingston Lowes is incalculably richer, and more subtle, than any connectionist explanation. For literary purposes, intuitive understanding (supported by sensitive scholarship) will do. It is an invaluable resource for psychological purposes too, not least as a challenge to future scientific explanation. But a theoretical account of mental association must go beyond intuition. It must give some idea, at least in general terms, of the fundamental mechanisms which underlie association and enable it to happen.

Connectionist ideas help us to comprehend how several passages about fishes or dolphins playing and gambolling in the waves might be associated in the memory. We can see why the memory need not be primed to look out for sea-creatures (why it can benefit from serendipity, discussed in Chapter 9). And we can see why, if it is so primed (by the poet's previous decision to write about a mariner), it is more probable that ideas of sea-creatures will be picked up.

We can suggest processes of pattern-completion that might enable a poet to recall a past context (a description of the luminous wakes of ships) from a single phrase ('in mari ludens', or 'playing in the sea') – or, for that matter, which might enable a novelist to remember a childhood experience on eating a madeleine. We can see how the mind might be sensitized to a word (such as 'hoary') because it appears in the context of something interesting (such as gambolling dolphins).

We need not imagine a mechanical dictionary-search for metre and rhyme to pair 'water-snakes' with 'hoary flakes': analogical pattern-matching would do. This process could find a rhyme-pair even if the best match, given multiple weak constraints of meaning and metre, was less acoustically exact than this one is. It could even allow such a match partly because of similarity in spelling ('love' and 'prove', perhaps).

We can see how probabilistic pattern-matching could meet a challenge mentioned in Chapter 1, selecting 'a summer's day' in preference to 'a winter's day' because of their different associations. We can see how two ideas could blend in the memory to give a third (as *go* and *went* blend so as to give *wented*). We can suggest a psychological mechanism to overlay one pattern on another without losing either, and to produce a new pattern to which both old ones (among others) contribute. We can even see how snakes of another kind – tail-biting snakes – might have been brought to life by Kekulé's reverie on 'gambolling' atoms, not gambolling dolphins.

In brief, we can now say something *specific* about how the hooks and eyes of memory might find each other, and how they might clip together.

You will have noticed the cautionary uses of 'might' in the preceding section. Connectionist research is in its infancy, and cannot explain all the unconscious operations linking literary sources with poetic imagery. Many more – and more powerful – types of connectionist system remain to be defined, and we cannot say now what their computational properties will be.

Moreover, current neural-network models, for all their likeness to the brain, are significantly unlike brains too. For instance, nearly all involve two-way connections, whereas brain-cells send messages in one direction only. Any one unit is directly connected to only a few others, whereas the lacy branches of a given neurone usually abut on many hundreds of cells. Computer models contain no analogue of the neurochemicals that diffuse widely through the brain. Further, neuroscientists still know very little in detail about what computations are carried out by brain-cells, and how. Even if we identified (by literary–psychological methods) the specific associations that gave birth to the water-snakes, brain-scientists could not tell us precisely how they occurred.

It follows that suggestive explanations, indicating how creative association 'might' be scientifically understood, are – as yet – all that can be offered.

Even so, the brainlike models described above may seem so impressive as to eclipse programs of the more familiar type. 'Surely,' someone might say, 'we can now see that traditional programs, designed to do heuristic problem-solving, are irrelevant to creativity.'

This conclusion would be mistaken. We cannot say 'Connectionism rules!', and leave it at that. Certainly, the brain is a connectionist system. But many people – including some leading connectionists – argue that to do conscious reasoning, or even to understand grammatical English, the brain may have to function *as though* it were a digital computer. That is, it may have to follow strict, hierarchical and even sequential rules – for instance, the rules of chess, grammar or arithmetic.

This is in principle possible: a connectionist machine can simulate one carrying out a sequence of symbolic transformations – and many connectionist researchers today are trying to do so. (The converse is also true; indeed, most current connectionist models are not implemented in connectionist hardware, but are simulated on digital computers.) The brain seems to have evolved the capacity to simulate a serial machine, a capacity which may be necessary for carrying out certain tasks. Von Neumann (digital) machines may be good at logic because only a von Neumann machine, or something functioning as though it were one, can do logic.

Many kinds of thinking, besides logic, require strict rules and carefully monitored sequential decisions. The last thing that chess-players want is a tolerant blending of the rules for pawns and rooks; nor would they welcome two moves being made simultaneously. The evaluation phase of scientific creativity typically calls for step-by-step thought. For example, even if the potential chemical significance of Kekulé's tail-biting snakes was glimpsed by means of association, it could be proved only by careful deduction and mathematical calculation.

Artistic creativity, likewise, often involves such reasoning. 'Anything goes!' is not a good motto for the arts. We can enjoy a disciplined integration of a Bach fugue and jazz-style, but we would not appreciate just any fuzzy mix of melodies. (Even two-headed men have to be drawn in a self-reflective way, as we saw in Chapter 4.) Deliberate thinking is involved when artists evaluate and correct their work: quite apart from the manuscript-corrections, Coleridge altered several words in successive published editions of *The Ancient Mariner*.

Moreover, conscious planning and problem-solving often goes on too. For instance, Livingston Lowes mentions Coleridge's starting one stanza with 'The Sun came up upon the left' and another soon after with 'The Sun now rose upon the right'. He suggests that Coleridge deliberately used the idea (found in his reading) of the sun's rising on different sides to solve a specific artistic problem: how to get the ancient mariner's ship around the world, to the Antarctic ice and thence to the baking doldrums of the Pacific, without detailing every part of the voyage. This problem arose because Coleridge had decided that the poem would follow the mariner on his travels, in such a way that one could trace the journey on a globe. But not every leg of the journey merited equal space.

(Stanley Kubrick solved an analogous aesthetic problem in his film *2001, A Space-Odyssey*, which began with a series of images representing the evolution of the solar system, and of life – and intelligence – on Earth. Kubrick leapfrogged over the entire history of technology by transforming the triumphant ape-man's twirling bone into a spinning space-ship.)

Coleridge's canny short-cut is a kind of poetic thinking for which the brain may need to simulate a von Neumann machine. So too is the poet's conception of the poem (or a composer's conception of a fugue) as an architectural whole. As Livingston Lowes put it, structuring a poem 'involves more than the spontaneous welling up of images from secret depths', for the form of a poem 'is the handiwork of choice, and a directing intelligence, and the sweat of a forging brain'.

This is not to say that we know just how this 'directing intelligence' should be modelled on a von Neumann machine, for we do not. It is difficult enough for a literary critic, or a poet, to give an intuitive indication of the sorts of thought-processes involved. It will be many years, if ever, before we can identify them in scientific terms. Even then, for reasons explored in Chapter 9, we would not be able to give a fully detailed reconstruction of the writing of *The Ancient Mariner.* But this does not destroy the main point: that poetic creativity requires a rich variety of mental processes, intelligible in (both connectionist and non-connectionist) computational terms.

In sum: we do not know exactly how Coleridge conjured his water-snakes, but we do have some promising scientific ideas to help discover the secret. The results of associative memory may be magical, as smiling unicorns or ghostly sailors are. But the computational processes involved are not.

7

Unromantic Artists

Enough of map-making! What about the maps? Granted that computational concepts are needed to explain creative thought, how have AI-workers actually modelled it? Do any of today's programs manage to depict creativity even as well as a twelfth-century *Mappa Mundi* depicts the globe? More to the point, what can their successes – and their many, many failures – teach us about creativity in people?

In discussing these questions, I shall focus first on programs concerned with the arts, and then on examples related to science. But this distinction is not clear-cut. Science requires empirical verification (and measurement, if possible), whereas art does not. Even so, the two domains share many processes in common.

For instance, analogy (discussed in this chapter) and induction (described in the next) are involved in both types of domain. Analogical thinking is common in science, as well as in art: think of William Harvey's comparison of the heart to a pump, or of Bohr's picture of the atom as a tiny solar system. Likewise, learning to recognize different styles of painting or music requires inductive thinking, much as learning to diagnose diseases does too.

Some computer programs have already produced valuable new ideas, as we shall see. Had a human mind thought of them, these ideas would have commanded respect, even admiration. One AI-program, an 'expert system' dealing with a particular area of biochemistry, has been listed as a co-author of a research-paper published in a scientific journal. Another was responsible for the idea behind a new scientific patent. And a third has produced novel art-works that are exhibited at galleries around the world.

It does not follow that the second Lovelace-question – whether computers could appear to be creative – must be answered with a 'Yes'. This question asks, in effect, whether computers can model creativity. And for modelling creativity, novelty is not enough.

Novel (and valuable) outputs, previously unknown to the programmer and perhaps to any human being, are undoubtedly intriguing – especially if the unaided human mind could not have produced them. They may even support the case for saying that a certain program is, apparently, creative. But they are neither necessary nor sufficient to make a program a strong candidate for creativity.

They are not necessary because, as explained in Chapter 3, psychological creativity (P-creativity) often produces ideas that are not historically novel (H-novel). They are not sufficient because, as we saw when discussing the geometry-program at the end of Chapter 5, a novel (and surprisingly elegant) output may have been generated in an uncreative way. Whether a program models creativity depends more on its inner workings than on the novelty-value of its outputs. The crucial question is whether the output was generated by processes that explore, test, map and/or transform the conceptual space inhabited by the program concerned.

Before discussing any examples, we must be clear about the point of the exercise. The point is not to answer the fourth Lovelace-question, whether computers can 'really' be creative. That question is not my prime concern, and it will be considered only in the final chapter.

The point, you might think, is to answer the second and third Lovelace-questions: whether programs can appear to be creative, and to recognize creativity. But even those Lovelace-questions are considered here only for the light they shed on the first: how computational ideas can help us to understand *our own* creativity. My aim is not to hand out accolades to computer programs, but to illuminate the ways in which human beings manage their originality.

These two chapters do not aim to set up a competition, pitting programs against people. If they did, we would win hands down. It will be abundantly clear that Nobel prizes, and the Prix Goncourt, are still safe for humanity.

For our purposes, current AI-programs are not mere puny rivals, to be pummelled without mercy or pushed hard on to the ropes. Nor are they impertinent imposters, to be relentlessly mocked. Rather, they are early scouts sent in to explore unfamiliar psychological territory. Their adventures – successful or not, 'humanlike' or not – help us to think clearly about our own minds.

Art may seem highly unpromising as a domain for computer creativity. Admittedly, computers are widely used by artists as tools, or even as imaginative aids. 'Computer music' employs sounds unlike those produced by orchestral instruments, and allows composers to experiment with computer-produced chords or phrases that they might not have thought of themselves. 'Computer graphics' (including computer animation) sometimes results in images of fascinating beauty, and allows human artists to create visual effects of novel kinds. And 'writing-programs' help both children and adults to plan and produce texts of a complexity and coherence which they could not have achieved without them.[1] But in almost all these cases, the human being is an essential (hands-on) part of the exercise: seeding, amending and sifting the output concerned.

Occasionally, the human is absent. For instance, images such as the Mandelbrot set, which has new types of internal structure on an infinite number of levels, are produced by purely automatic means (Figure 7.1). In glorious Technicolor, they are glorious indeed – and have been exhibited in art galleries, accordingly. But they have a coldly mechanical aspect. More to the point, the computational processes involved are so unlike human thinking that they are of little psychological interest – except insofar as they show that unexpected

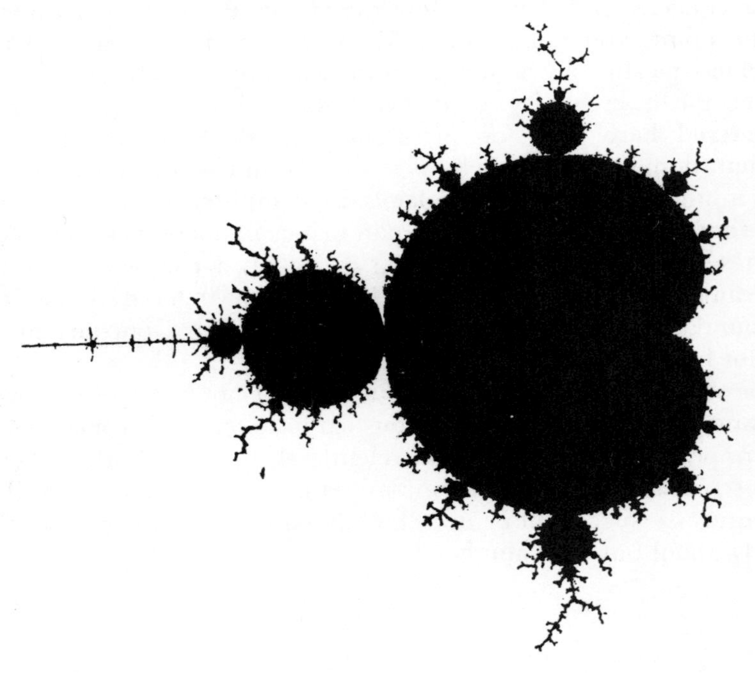

Figure 7.1

complexities may arise from very simple processes. (The Mandelbrot set is generated by the repeated calculation of one simple mathematical formula, $z \rightarrow z^2 + c$, the results of one calculation being used as input to the next.)

Our concern is not with 'inhuman' programs like these, but with programs specifically developed to cast light on the creativity of human artists: musicians, painters, poets or novelists.

Among the most successful of such programs so far are a series of drawing-programs written by Harold Cohen.[2] Cohen was already a well-established painter, with canvases exhibited in the Tate Gallery and many other museums, before turning to programmed art. But 'turning' is perhaps the wrong word here, for his current interests are, with hindsight, a nigh-inevitable development of his artistic career.

Cohen's own paintings were abstract, in the sense that they did not depict recognizable things (as in a still-life) or even fantasy objects (as in a Bosch or a Dali). However, people – himself included – interpreted them as representations of neighbouring and overlapping surfaces or solid objects, and he became deeply interested in the cognitive processes involved in such interpretation and representation.

He continually produced new variations, and new styles, seeking some general understanding of our emotional and perceptual responses to them. For instance, he investigated the differences in response to open and closed curves, and to symmetry and shading of various types. In other words, he systematically explored the conceptual space generated in our minds by the interaction of line, form and colour. This long-standing preoccupation with the psychology of art was the main reason for his later interest in computer-generated canvases.

A second reason was Cohen's growing conviction that art is, to a large extent, rule-governed. He had experimented with various 'rules' for painting, so that (for example) a line would be continued in a direction determined by some pre-existing, and sometimes largely random, feature.

Shortly before turning to programmed art, he said (in a BBC interview): 'I think, at each stage in the painting, I am placed in a new situation where I have to make a decision in relation to what's already been done. . . .' And (in a statement to the Arnolfini Gallery) he explained: 'I'd always reach the point in a painting where it was a question of saying: Well should I make it red or yellow? . . . I wanted to arrive at the state where the colour was as unequivocal, as positive, as

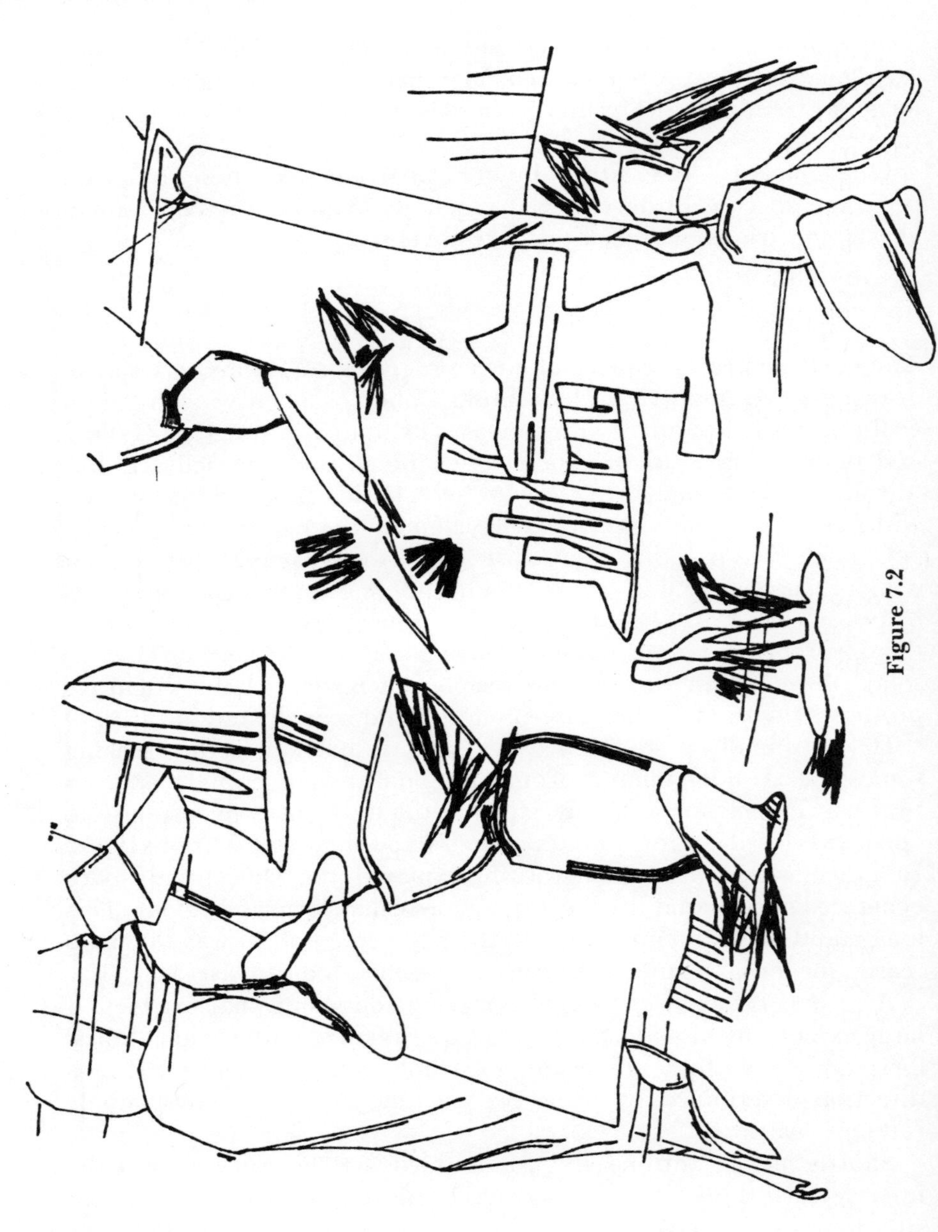

Figure 7.2

unarbitrary, as the drawing.' (Significantly, he remarked a few years later that a program for generating maze-like structures 'had the interesting result . . . of blocking my longstanding preoccupation with colour, since I could find in it no rational basis for a colour organization.')[3]

Small wonder, then, that Cohen embarked on a voyage of computational discovery. What is perhaps more surprising is the aesthetic quality of the results. Cohen's computer-generated drawings are shown and commissioned around the world – and not only for their curiosity-value. The Tate Gallery, for instance, mounted a special exhibition in 1983 of his abstract designs resembling landscapes (see Figure 7.2).

Cohen's programs are continually developed, to cope with added aesthetic complexity and an increasing range of subject-matter. For example, the ANIMS program that drew the bulls in Figure 7.3 was later able to draw Figures 7.4 and 7.5. (Cohen remarks that Figure 7.3 is in certain ways similar to African Bushman and Australian Aborigine art, whereas Figure 7.5 is more like the Altamira and Lascaux cave-drawings of the Northern Palaeolithic.)

His most interesting computational project so far, both aesthetically and psychologically, is a program – or rather a series of programs – called AARON.

The development from one version of AARON to the next can involve a fundamental change in the program's nature, a radical alteration in the conceptual space it inhabits.

The early AARON of Figure 7.2 concentrated on spontaneous drawings of abstract forms which (in the viewer's eye) could sometimes be seen as rocks and sticks scattered on the ground, or occasionally as strange birds or beetles. Human figures were not even dreamt of in its philosophy. By contrast, the frontispiece shows a picture deliberately drawn by a maturer version (of 1985), whose aesthetic world contains far more challenging creatures. (As remarked in Chapter 1, I keep this drawing in my office, where it has been innocently admired by many visitors unaware of its origin.)

Later, AARON's drawings became more complex still, depicting groups of human figures in a jungle of vegetation (an example from 1987 is shown in Figure 7.6). The program's most recent images (Figure 7.7, drawn in 1989) show human figures of a fully three-dimensional kind, as opposed to the two-and-a-half dimensions of the frontispiece acrobats.

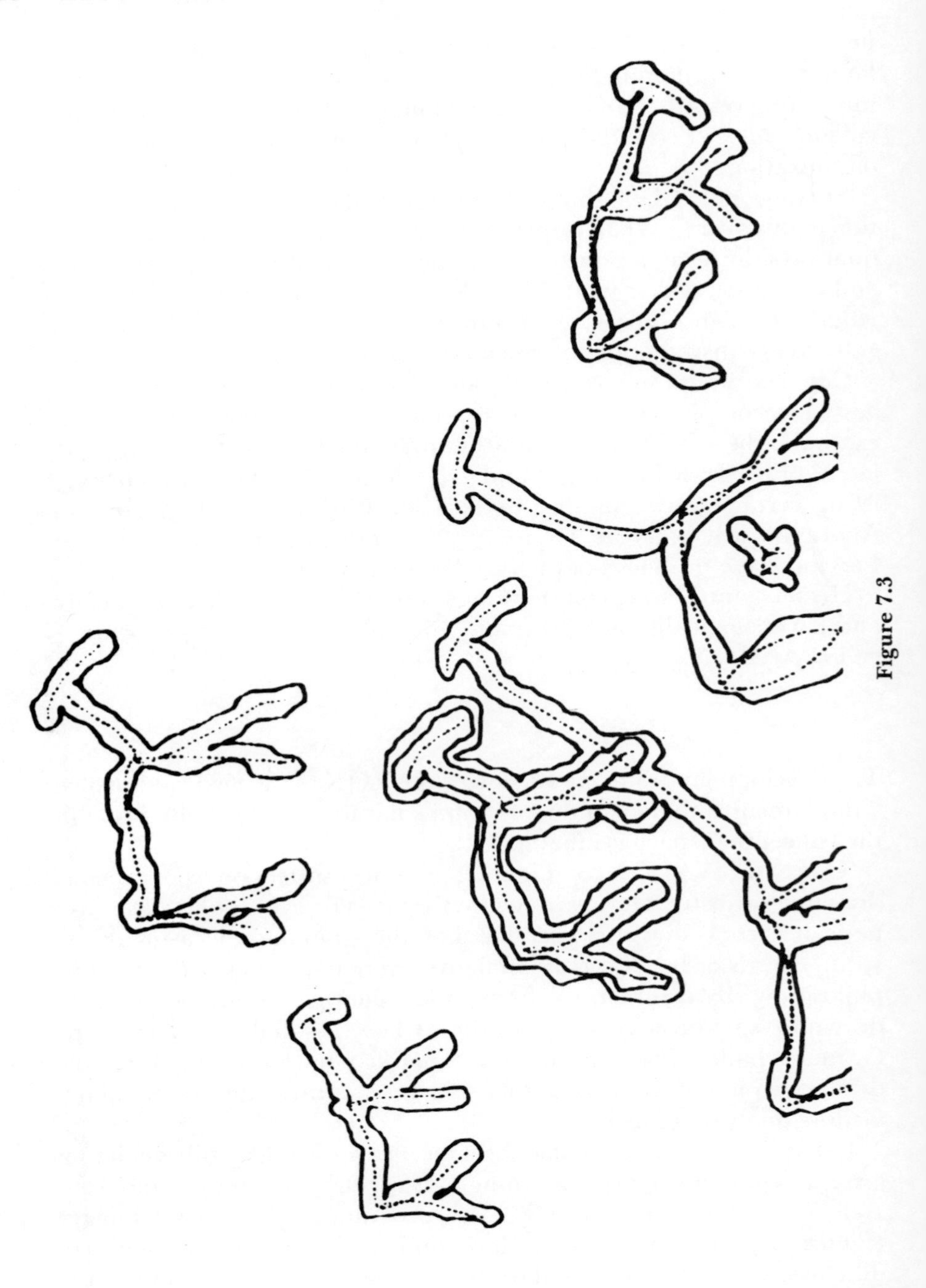

Figure 7.3

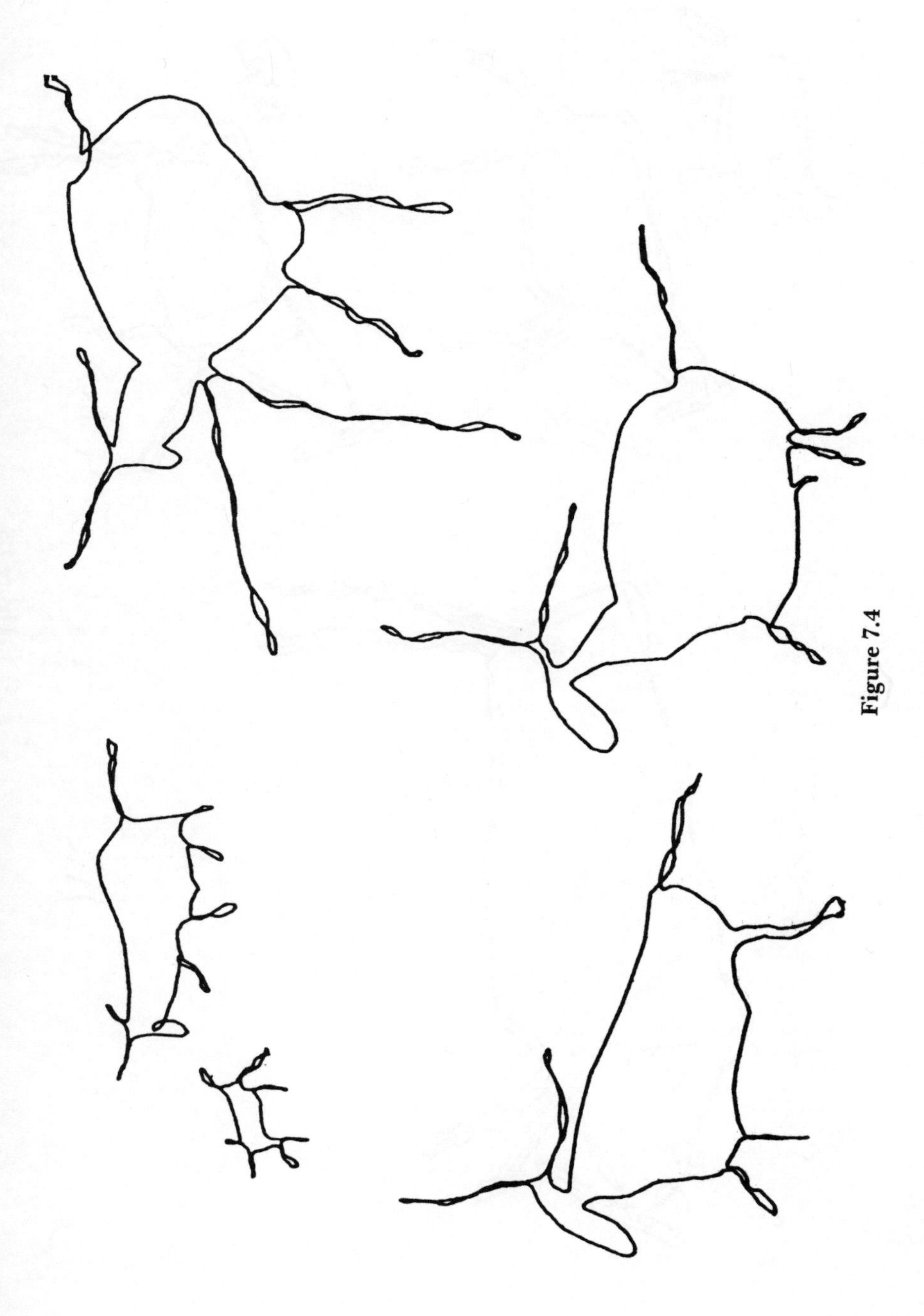

Figure 7.4

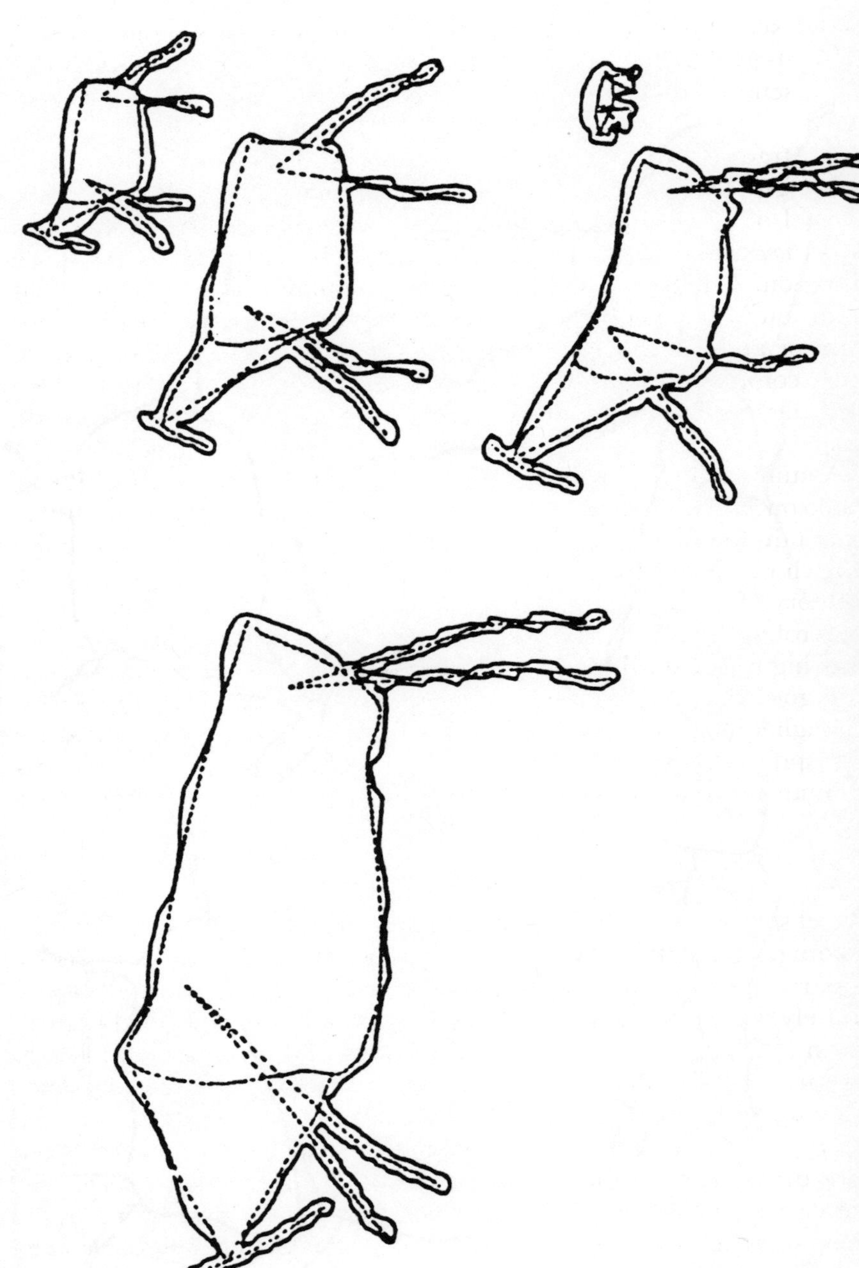

Figure 7.5

I described abstract-AARON's drawings as 'spontaneous' and acrobat-AARON's as 'deliberate' because only acrobat-AARON can purposefully consider what sort of picture it is going to draw before it starts.

Abstract-AARON draws its landscapes by randomly choosing a starting-point somewhere on the paper, and then continuing under the control of a collection of IF–THEN rules (what AI-workers call a *production system*) which specify what should be done next in any given situation. Whether a line should be continued, and if so in what direction, will depend for example on whether it forms part of a closed form or an open one. Abstract-AARON's IF–THEN rules may be fairly complex. Several aspects (not just one) of the current state of the drawing may have to be checked, before the program knows what to do next.

Assuming the rules to be sensible ones, the program can thus be relied on to make locally coherent decisions (including random activity at certain specified points). But it cannot explicitly consider the picture as a whole. (Nor can it learn from its own past actions, because it has no memory of them.)

Acrobat-AARON, by contrast, can plan certain aspects of its drawing before putting pen to paper. And it can monitor its execution as it goes along, to check that the planned constraints are being met. Though no less autonomous than abstract-AARON, it is in that sense less spontaneous. (And in that sense, likewise, van Eyck was less spontaneous than Jackson Pollock.)

To get some sense of the conceptual spaces involved here, consider how you might go about drawing a picture to be called 'Acrobats and Balls'. A particular content and composition must be chosen, and appropriately executed. But before putting pencil (or pen, or Japanese brush . . .) to paper, the overall artistic style has to be decided. Let us assume that we want a sketchy, non-surrealistic, pen-and-ink line-drawing of the kind shown in the frontispiece.

Try, for a while, to draw a picture in that style. Even if you are a very poor draughtsman, and are reduced to copying the frontispiece, the exercise should be illuminating. Then, spend some time trying to jot down some advice (a list of dos and don'ts) which might help a friend to draw such a picture – preferably, a friend who has not yet seen any of Cohen's work.

What sorts of advice might be included on your list? Your hints on

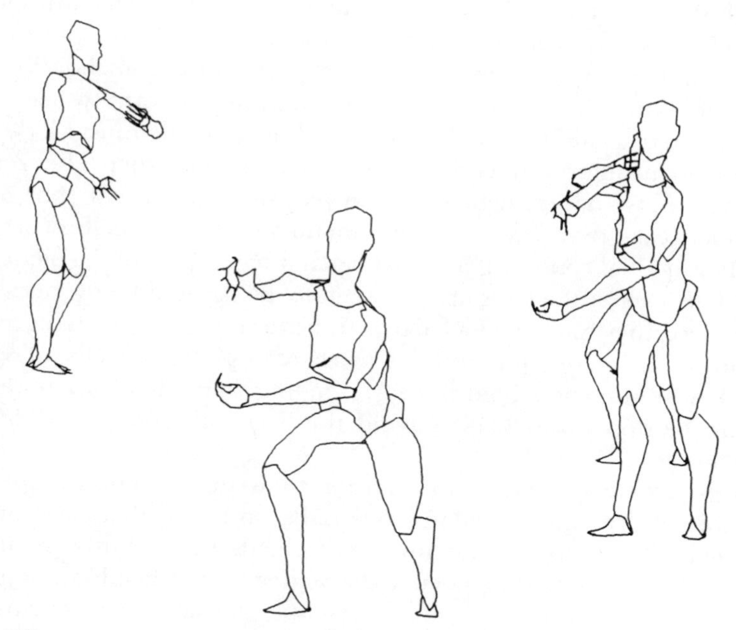

Figure 7.7

execution will not include any positive guidance on shading, since no shading is allowed (although a very small amount of 'hatching' is permitted). But you might point out that, if solid objects are to look solid, your friend must represent occlusion by interrupting the outline of objects lying behind other (nontransparent) objects. (There is no line in the frontispiece depicting the convex surface of the middle acrobat's left knee, because the left wrist and hand are in front of it.)

The body-outlines being the aesthetic focus, there are only minimal (perhaps even optional) indications of the clothes and facial details. Noses appear to be *de rigueur*, but lines suggesting footwear are not. As for the body-shapes, these need not be so lifelike as they would be in a Baroque cartoon. But they must be fairly realistic: no late-Picasso faces with both eyes on one side of the nose, or late-Picasso limbs depicted as conjoined wedges or triangles.

There can be no gravity-defying postures either: no human figures floating horizontally in the air, as in Chagall's dreamscapes. The balls must obey the laws of gravity, too. And their position – whether in the air, on the ground or somehow supported by the acrobats – must be plausible with respect to the bodily attitudes of the human figures. These attitudes are represented not only by the angle and position of the limb-parts, but by their foreshortening and/or bulging muscles (see the left upper-arm of the balancing acrobat in the frontispiece).

As for the composition, the ground-plane must somehow be made evident; but there should be no horizon-line, nor any other explicit indication of ground-level. The 'space' of the drawing must fill the page: no trio of tiny figures squeezed into the top lefthand corner. The picture as a whole must be aesthetically balanced, or symmetrical (but not too symmetrical – that way lies the inhumanity of the Mandelbrot set). And each individual content-item must be an integral part of the picture. For example, even if the balls are not actually being used by the acrobats in some way, they must not look like extraneous litter strewn on the ground.

One could go on, and on . . . Clearly, completing this list of dos and don'ts would be no trivial task. And telling your friend *how* to do what has to be done, and *how* to avoid the listed pitfalls, would be more difficult still.

It is orders of magnitude harder to write a computer program capable of drawing 'Acrobats and Balls' in an aesthetically pleasing manner. The programmer must not cheat, by building in just one picture line by line (like a poem to be recited parrot-fashion), or just one compositional form. Nor can he escape the charge of cheating by building in twenty pictures, or seven compositional forms.

Rather, the program should be able to generate indefinitely many pictures of the same general type. It should continually surprise us, by drawing pictures it has never produced before. And, although some of these drawings will be less satisfactory than others, only a few should be unacceptably clumsy – and none should be aesthetically empty.

This is what Cohen has achieved. All the versions of AARON can draw new pictures at the touch of a button. At the close of the world-fair in Tsukuba (Japan), the organizers sent Cohen the 7,000 drawings which the program had done there: each was unique, and none had been seen by him before. Moreover, they have brought pleasure to many people around the world. For Cohen's program produces aesthetically satisfying results. It is not like a dog walking on its hind legs, of whom Dr Johnson said 'The wonder is not that it does it well, but that it does it at all.'

Admittedly, it is not Leonardo. And, admittedly, I have chosen one of my favourites as the frontispiece. Not all of AARON's acrobat-pictures are quite as attractive as this one; the two compositions shown in Figures 7.8 and 7.9, in my view, are not. But virtually all are pleasing, and many are spontaneously praised by people who do not know – and might be loath to believe – that they were generated by a computer program.

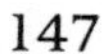

Figure 7.8

What's the secret? How does AARON do it? The answer, as in most cases of creativity, lies in a mix of general and specific knowledge. The program's grasp of both types of knowledge has improved with successive versions, but the change in specific knowledge is more immediately apparent.

Abstract-AARON, for example, knew nothing about human figures and so was incapable of drawing them. By contrast, the later acrobat-AARON was provided with an underlying model of the human body.

Unlike the articulated wooden models sometimes used in artists' studios, this 'model' is purely computational. It consists not of a list of facts (or 'knowledge that'), but of a set of processes (or 'knowledge how').

AARON's body-model is a hierarchically structured procedural schema, defining a search-tree which (with varying details slotted in at the decision-points) can generate an indefinite number of line-drawings representing a wide range of bodily positions. That is, it

Figure 7.9

constitutes a conceptual space (or a family of neighbouring conceptual spaces), whose potential is both created and limited by the constraints concerned.

Some constraints are inescapable: for AARON, all bodies have two arms and two legs. The program cannot represent a one-armed acrobat (no 'funny men' here). Certainly, an acrobat's right arm need not be depicted in the final execution of the sketch; it may be hidden behind someone's back, for instance. But AARON's original conception of the picture would have featured it. That is, its body-model tells it only how to draw two-armed people. One-armed people are simply not allowed for.

To be able to represent one-armed acrobats, AARON's body-model would have to resemble a computational frame (as described in Chapter 5) with a slot for 'number-of-arms', whose default-value would be 'two' but which could accept 'one' – and perhaps even 'zero' – as well. (Suppose Cohen were to provide AARON with this slot-filling mechanism: could the program then go merrily ahead, reliably producing pretty pictures with *just this one* amendment? Or would some other alterations have to be made, too?)

Other constraints defined within AARON's body-schema are context-sensitive. For instance, the nose establishes which way the head is turned – which is why noses are *de rigueur* in the pictures drawn by acrobat-AARON. (The heads in Figure 7.7 do not all have noses: the body-model underlying this picture involves a richer knowledge of three-dimensional space, so 'directional noses' are no longer needed.) Likewise, whether – and how – the limbs are foreshortened, or the muscles visibly flexed, depends on the specific bodily attitude and/or viewpoint.

Whether the limb itself is flexed depends on a number of things also. The joints, and their freedom of movement, are represented in the program's body-schema. But the attitude of a leg varies according to whether it is providing the body's sole support, and whether it has the ground or a curved surface beneath it. If it is not the sole support, then AARON must know what other object is involved (compare the legs of the two frontispiece acrobats semi-supported by balls). Likewise, the attitude of an arm depends on whether it is supporting, throwing or catching a ball, and whether it is being used to balance the acrobat's body. (If a 'one-arm' possibility were to be added, as suggested above, Cohen would have to modify the heuristics dealing with balance.)

Another consideration that affects AARON's decisions about the attitude of the limbs is the overall composition. It is not simply because of the constraints involved in body-balance that the third (pirouetting) acrobat has the left arm raised and the right arm extended.

To produce paintings as well as drawings, AARON would need to know about colour. However, the aesthetics of colour are very subtle – as human painters already know. Jungle-AARON, which drew Figure 7.6, would presumably require knowledge of individual colour-facts, such as that – even in the tropics – the ground is not sugar-pink (a fact sometimes deliberately ignored by Gauguin). But even abstract-AARON would need some grasp of the general principles governing the aesthetics of neighbouring colours in a painting.

Because these principles are so obscure, Cohen did not try to make his early drawing-programs do colouring, too. (The coloured versions of AARON's designs that are exhibited were hand-painted by Cohen himself.) He is currently working on the problem, and a prototype colouring-machine is about to be tested (in summer 1990). To the extent that it succeeds, AARON will become a painter as well as a draughtsman.

The general knowledge which AARON already possesses concerns such matters as how to represent occlusion, solidity or illumination in a line-drawing. This might seem fairly simple. Occlusion, for example, requires the artist to interrupt lines depicting surface-edges of the occluded object. But which lines, and at just which points should the interruption start and finish? To answer these detailed questions about execution, the program must choose a specific viewpoint, to be kept consistent across all the objects in the drawing.

Similarly, solidity and illumination are represented by hatching close to an edge-line. But which edge-lines? In Figure 7.2, a direction of illumination is suggested by the fact that all the hatching lies on the same side of the closed forms concerned; but in the frontispiece, the arms of the sitting acrobat are hatched on opposite sides.

To be sure, these are two very different sorts of hatching: perhaps if the acrobat's inner arms were both hatched more heavily, the result would be a visual absurdity? Yes, perhaps. (And perhaps Cohen has already tried it.) This is just the kind of exploratory question that is typical of creative thinking.

So, too, is the question of how the artist can indicate the ground-plane – either with or without a horizon-line. Perhaps it is only prior knowledge about the anatomical stability, and the relative sizes, of human beings which enables acrobat-AARON, and the viewer, to interpret the frontispiece sensibly. (The young Picasso's circus-pictures include some showing child-acrobats; how could AARON draw a recognizable child-acrobat without confusing the ground-plane?) In the jungle-drawing of Figure 7.6, the ground-plane is indicated by strewing rocks on it.

A third exploratory question concerns the nature of aesthetic symmetry, or compositional balance. The frontispiece would look very strange without the pirouetting acrobat, and abstract designs have compositional constraints just as 'realistic' ones do.

Cohen had to consider all these questions in order to write his programs. 'Three cheers for Cohen!', you may cry. There is no doubt that Cohen himself is creative. Moreover, his programs help us to think clearly about some of the mental processes involved when a human artist sketches abstract forms, or acrobats – and that, after all, is our main concern. But are they themselves strong candidates for creativity?

Well, what must a program be like, to appear creative? For a start, it must inhabit, and explore, a conceptual space rich enough to yield indefinitely many surprises. Ideally, it should extend this space – or perhaps even break out of it, and construct another one.

It must produce P-novel, and preferably H-novel, results. The results must often be individually unpredictable, although they may all possess a recognizable conceptual style. They must be generated by the program acting alone, relying on its own computational resources rather than constant input from a human operator (this does not preclude specific 'commissions', such as 'Please draw two acrobats with one ball').

Further, the program's computation must involve judgment. Purposeful behaviour should be more common than random processes, and any randomness must be constrained by the general nature of the creative domain concerned. Preferably, the program should sometimes be able to reconsider its past choices in deciding what to do next (although, as we shall see when discussing jazz-improvisation, this self-critical ability is not always available even to human creators).

The program's newly generated structures must be recognized by us as valuable in some way: in AARON's case, as being aesthetically pleasing. It must have a way of evaluating various possible structures for itself, so that it can avoid nonsense – and, ideally, cliché. (If it lapses occasionally, it can be forgiven: what human artist or scientist does not?)

If the relation between the program's generative strategies and its results casts light on human creativity, not forgetting the creativity of the audience in interpreting the novel ideas, so much the better. Indeed, from our point of view this is what matters: we are interested in 'creative' computer programs only to the extent that they illuminate human psychology.

AARON meets all these criteria of creativity – with one important qualification. Since each of its newly minted drawings is unique, each is a historical novelty. But only if we think of the various versions of AARON as a *single* computational system do its drawings count as H-novel (or even as P-novel) in the strong sense defined in Chapter 3. With reference to abstract-AARON's abilities, the drawings of acrobat-AARON *could not* have happened before; likewise, acrobat-AARON *could not* have drawn the jungle-scene in Figure 7.6.

If we consider only a particular version of the program, however, the matter lies differently: each drawing *could* have been generated (by that version) before. Moreover, the progression from abstract-AARON to acrobat-AARON, and thence to jungle-AARON, is due directly to Cohen, not to any autonomous self-modification by the program.

In short, AARON's originality is not truly radical, because its exploration of its own conceptual space is relatively unadventurous. The creativity of any given version of AARON is more like that of the child who repeatedly says, 'Let's make another necklace, with a different number of beads in it,' than of the child (perhaps the same one, an hour later) who says, 'I'm bored with addition-by-necklace!

Let's do subtraction now.' The program does not seek to test the limits of its creativity, as the child does (by trying to do '1 + 1 + 1', or to make a necklace with just seventeen beads). Since it cannot test its constraints in these ways, it is hardly surprising that it does not try to change them either.

AARON is like a human artist who has found a style, and is sticking to it. This achievement is not to be scorned: the favoured style may allow the creation of many different pictures (or poems, or melodies), each one attractive in itself. But adventurous, it is not.

Ideally, a creative drawing-program should be able to switch to a new content and/or generate a new style. It should be able, like Picasso perhaps, to think: 'I'm bored with acrobats! I'll draw Minotaurs instead. And I'd like to explore a different style: I'll try drawing limb-parts as straight-sided geometrical figures, and see what happens.'

For this to be possible, the program would need the relevant types of knowledge. Also, it would need a way of reflecting on that knowledge, so as to be able to describe, compare, criticize and alter it. In other words, it must be able to construct, inspect and change various maps of its mind.

With respect to knowledge of content, no one can draw the Minotaur without knowing what it looks like. This means knowing, among other things, how it supports itself and how it can move. If Cohen had also provided body-schemata representing Minotaurs, AARON could have varied the content of its pictures from time to time.

Human artists can draw other things besides, and can imagine content-items they have never seen – such as unicorns and water-snakes. But, unlike AARON, they have the benefit of years of experience, in which to garner richly associated representations of many different things. Unlike AARON, too, they can associate visual with symbolic content – hence the mythological significance of the Minotaur, and the unicorn's sweet expression.

As for the knowledge required to change an artistic style, AARON would need (for instance) to be able to see the visual analogy between a straight line and a gentle curve, and therefore between a thigh and a wedge or a triangle. (Many humans cannot see this analogy until someone like Picasso points it out to them, and even then they may resist it.) It would be relatively easy to provide AARON with this systematic deformation of curves into lines, but more difficult to enable the program to generate it (and other stylistic variations) unprompted.

Cohen is well aware of these limitations on AARON's creativity. His ultimate aim is to produce an AARON that can modify the way in which it does its drawings. It will be easier to provide for superficial variations of a given style than for the creation of a new style altogether. To break out of one style into another, one must (without losing overall coherence) modify one's generative procedures at a relatively fundamental level.

For either sort of stylistic change to happen, self-criticism is essential. AARON has no way of reflecting on, and transforming, its own activities. As yet, only a few programs can do so. These systems, as we shall see in Chapter 8, have heuristics for modifying conceptual spaces in promising ways, including heuristics for changing their own heuristics.

In principle, then, some future version of AARON might autonomously do a pleasing drawing which it *could not* have done before. No matter if the drawing were less beautiful than a Leonardo, and less surprising than a Matisse: creativity is a matter of degree (and most human artists cannot reach such heights). AARON, in that event, would have done all that can reasonably be asked of a creative program.

Some people, of course, would still refuse to call AARON creative. They might seek to justify their rejection in terms of the intrinsic aesthetic interest – or lack of it – of the program's drawings.

Suppose, for example, that AARON were to draw acrobats with triangular calves and wedge-shaped thighs, and that no human artist had done so before. Instead of treating the drawing in an open-minded way, some sceptics would stubbornly reject them.

'Yes,' they might be forced to admit, 'there is some analogy between limb-parts and wedges. But it is wholly uninteresting, and drawings based on it are downright ugly.' And this prejudice would persist. In their view, it is one thing to allow a human artist to challenge our perceptions, and upset our comfortable aesthetic conventions, but quite another to tolerate such impertinence from a computer program.

At base, this attitude has nothing whatever to do with the intrinsic nature of the program's drawings. Rather, it springs from the assumption that no program can *really* be creative, no matter what novelties it manages to produce. This assumption, as we shall see in Chapter 11, is not a factual belief about psychology but (in large part) a moral attitude. As such, it does not affect the (factual) questions whether programs might appear to be creative, and whether they might throw light on human creativity.

In sum, AARON's performance – like *any* computer-performance –

is in principle irrelevant to the fourth Lovelace-question. But it gives us good reason to answer 'Yes' to each of the first three Lovelace-questions. Future versions should give us better reason still.

What of music? Can computer-music help us to understand our own musical abilities? Our interest here is in music written *by* computers, not *with* them. Are there any programs that generate new pieces from start to finish – and if so, are they convincing candidates for creativity?

There is no computer-generated 'Beethoven's Tenth'. But there are some music-writing programs focused on less exalted genres, such as nursery-rhyme melodies and modern jazz. Nursery-rhymes, you may feel, are not worthy of your attention. But jazz is certainly complex enough to be aesthetically interesting. How does a jazz musician create pleasing harmonies, interesting rhythms and acceptable melodies?

These questions have been tackled by the psychologist Philip Johnson-Laird.[4] An accomplished jazz-pianist himself, he has written a program modelling improvisation in the style pioneered by Charlie Parker and Dizzy Gillespie. The program generates schematic chord-sequences and then, using the chord-sequence as input, improvises both actual chords and bass-line melodies (and rhythms).

Modern jazz is a special case of tonal music. It is based on underlying chord-sequences (the twelve-bar blues, for example), agreed before the performance is begun and repeated until (after a certain number of repetitions) it ends.

Each chord in the chord-sequence is defined by its key and its harmonic type (tonic, dominant and so on); six types are commonly used. Modulation is more common, and more free, within chord-sequences than in classical music. For instance, the jazz-musician can modulate to *any* new tonic (although Johnson-Laird mentions one harmonic interval which is very rarely used).

The conceptual space of all possible chord-sequences is not only very large, but structurally complex. Modulations, for instance, involve the embedding of mini-sequences within other sequences – whose beginning and end have to 'match' (for example, by returning to the home key).

Johnson-Laird points out that this hierarchical musical space can be defined only by a grammar of considerable computational power. 'Nesting' chord-sequences within sequences, while keeping the harmonic context coherent at every level, puts a large load on memory – so large, that a written notation (allowing for deliberate backtracking) is helpful. Indeed, jazz-composers may take many hours of careful thought to build their chord-sequences, which they write down using a special harmonic notation. (This is an instance of the general point made in Chapter 5: that new representational systems can make new kinds of thought possible.)

To appreciate what Johnson-Laird is saying, we need not wrestle with the jazz-buff's arcane musical notation. (It is even less familiar to most of us than are bar-lines, crotchets and quavers.) Instead, consider an analogy from language: *The potato that the rat that the cat that the flea bit chased around the block on the first fine Tuesday in May nibbled is rotting*.

If you can understand this multiply nested sentence without pencilling in the phrase-boundaries, or at least pointing to them, I take my hat off to you. If someone were to read it aloud, without a very exaggerated intonation, it would be unintelligible. Moreover, you would find it difficult, perhaps impossible, to invent such a sentence without writing it down. For you cannot select the word *is* without remembering *potato*, twenty-two words before. (If you had started with *The potatos* . . . you would have needed *are* instead.)

Johnson-Laird has written a program that first generates a simple chord-sequence (compare *The potato is rotting*), and then complicates and embellishes it to produce a more complex one (comparable to the longer sentence given above). Every modulation and embellishment is constrained by the relevant harmonic context (much as the verb *is* is constrained by the far-distant noun, *potato*), and whenever the rules allow several possibilities, the program chooses at random.

The reason why ordinary musical notation (staves, clefs and crotchets) cannot be used to represent chord-sequences in jazz is that the 'chords' in chord-sequences are really *classes* of chords. That is, any chord (so defined) can be played in many different ways – and one aspect of jazz-improvisation is to decide exactly how to play it.

Usually, there is at least one occurrence of four specific notes, including the root. But each of these notes may vary in pitch; thus the chord of C major may be played using middle-C, top-C or both. Occasionally, the root is omitted. Other notes are sometimes added (usually the 6th or the 9th). And the chord can be inverted, so that the root is the note of highest pitch. The space of possibilities, then, is very large indeed.

Here, however, we are not in the land of the rat-nibbled potato. For jazz-musicians decide how to play each chord 'on the fly', without consciously thinking about it. This suggests, says Johnson-Laird, that the generative grammar they are using is a relatively simple one, which puts very little strain on working memory. If they continually had to do the musical equivalent of composing the monstrous sentence cited above, they would never be able to improvise so effortlessly.

Computer memories, of course, are less limited. But Johnson-Laird – like me, and perhaps you too – is interested in human minds, not computers. So he does not cheat, by taking advantage of the machine's inhumanly efficient memory to enable his jazz-program to decide how to play a particular chord. Instead, his program makes its decision by using harmonic rules that do not refer to any chord prior to the previous one (much as if you could have selected *is* by reference not to *potato* but to *nibbled*).

What about the melodic constraints? 'Easy!' you may say: 'They are known already. My local music-shop sells books describing them.' Not so. What your music-shop sells is lists of phrases, or motifs, to be memorized and strung together (suitably harmonized).

Although some jazz-writing programs work in this way, Johnson-Laird's does not. The reason is that he is modelling human psychology.

To try to improvise melodies like this would tax our long-term memory beyond its limits (there are simply too many different motifs for us to remember). Moreover, many melodies are played only once, and our musical intuition – our tacit grasp of the conceptual space concerned – tells us that many more, which will never be played, are *possible*. (Just as one can map a country without actually visiting every part of it, so one can define a conceptual space without generating all the instances that lie within it.) Above all, somebody had to write the motifs in the first place. A program intended as a psychological theory of jazz-improvisation should be able to explain how they did so.

Melody, too, is improvised on the fly by human musicians. So Johnson-Laird uses a grammar with relatively limited computational power to generate the bass-line. Once his metrical program, which generates jazz-rhythms, has determined the timing (and intensity) of the next note, the melodic program chooses the note's pitch. Its choice is guided partly by harmony, and partly by some remarkably simple constraints on melodic contours (the rise and fall of pitch).

It has been known for some time that aesthetically pleasing melodies tend to involve a succession of small intervals followed by a larger one, and *vice versa*. And a musicologist's 'directory' of themes in classical music claimed to identify melodies *uniquely* by describing their first fifteen (or even fewer) notes in terms merely of rising or falling pitch.[5] For example (using * for the first note, R for repeat, U for up, and D for down), the opening of Beethoven's Fifth has the contour *-R-R-D-U-R-R-D, and 'Greensleeves' opens with *-U-U-U-U-D-D-D-D. Having checked that these ideas fit a large sample of jazz bass-lines too, Johnson-Laird adapted them to write a simple grammar for producing melodic contours.

The grammar forbids contours with a succession of alternating large and small intervals, and it suggests a large jump after a series of small ones (and *vice versa*). It can give four instructions, some of which (like adjectives in Dickens' sentences) can be repeated several times. These are: first note; repeat of previous note; small interval; and large interval.

The rules do not distinguish between rising and falling pitch, because whenever Johnson-Laird inverted a classical theme coded by rise-and-fall, he ended up with another theme in the musicologist's directory. In other words, 'rise' and 'fall' seem to be interchangeable at will, if the task is to decide whether a series of notes is a *possible* melody (as opposed to deciding which melody it is).

The melody itself is created on-line by the jazz-program. A melodic contour is generated by the contour-grammar as the melody develops. If the current step of the contour instructs the program to 'repeat' the previous note, then there is no more to be said. But suppose the contour tells the program to pick a 'small interval'. In that case, harmonic constraints (the underlying chord, and rules about what passing-notes may be used between chords) decide which particular interval this can be. If several intervals are allowed, then one is chosen at random.

Does the jazz-improviser appear to be creative? Well, Johnson-Laird describes its overall performance as that of 'a moderately competent beginner'. So the criterion of 'positive evaluation' is satisfied, to some degree. After all, many people would be happy, at least for a while, to have a moderately competent beginner playing jazz on their living-room piano.

Moreover, the program explores an interestingly complex musical space. It comes up with P-novel – and possibly H-novel – chord-sequences and melodies, which are unpredictable in detail because of the random choices involved.

Like Cohen's drawing-programs, however, it can create only within a recognizable artistic style. It has no rules capable of transforming (for example) the lower-level rules that generate chord-sequences. Consequently, it never produces any music which it *could not* have produced before. But then, how many jazz-musicians do?

Over thirty years ago, a program was co-writing poems in the form of Japanese haiku, a highly economical genre that constrains the poet as much as possible.[6] The program contained a three-line frame, with nine slots for which the human operator would choose words:

All [1] in the [2]
I [3] [4] [5] in the [6]
[7] the [8] has [9]

Sometimes, the human operator was given a free choice: any word could be chosen for any slot. But sometimes the program constrained the choice.

The haiku-program contained a simple semantic net, built on the principles of a thesaurus. The available words (about one hundred and forty in all) were grouped into nine separate lists, each numbered for the relevant slot. The list for slot-7 contained eleven words (including *Bang*, *Hush*, *Pffft*, *Whirr* and *Look*), while the list for slot-5 contained twenty-three words (such as *Trees*, *Peaks*, *Streams*, *Specks*, *Stars*, *Pools* and *Trails*). A few words appeared on two different lists (thus *White* could fill slot-1 and slot-4).

When choosing a word for slot-n, the human operator was forced to go to list-n. Moreover, each slot in the haiku-frame was semantically connected (by one or more links) to another slot (or slots). The semantically central word was slot-5, which was directly linked to five others, and indirectly linked to the rest. The operator had to obey these semantic constraints also, when choosing words from the lists.

This man-machine procedure produced the following two examples:

All green in the leaves
I smell dark pools in the trees
Crash the moon has fled

All white in the buds
I flash snow peaks in the spring
Bang the sun has fogged

For comparison, here are two haikus of the same form where the human operator was allowed to choose words freely:

Eons deep in the ice
I paint all time in a whorl
Bang the sludge has cracked

Eons deep in the ice
I see gelled time in a whorl
Pffft the sludge has cracked

Can you see a significant aesthetic difference? I cannot. I see no evidence that the human's creativity is diminished by having to follow explicit semantic constraints, instead of implicit ones. (The operator breaks a rule, to be sure, in omitting *All*; but only seven of his freely chosen words are foreign to the program's list, and he seems to be temporarily fixated on *Eons deep in the ice*.)

The reason for the apparent success of this very early program lies less in the program itself than in its audience. In other words, readers of this particular literary form are prepared to do considerable interpretative work. The program's gnomic poems are acceptable to us accordingly.

In general, the more the audience is prepared to contribute in responding to a work of art, the more chance there is that a computer's performance (or, for that matter, a human artist's) may be acknowledged as aesthetically valuable. We have already seen this process at work with respect to abstract-AARON's 'landscapes'. The scare-quotes are there because it is the viewer, not the program, who gives its designs this interpretation. Much of the beauty, one might say, lies in the eye of the beholder.

Certainly, the beholder is sometimes the program (or human artist) itself, functioning in the self-reflective 'evaluative' mode. But the audience may be prepared to supply meaning which the creator does not – or even cannot. (This fact is stressed in much recent literary theory, wherein the text alone, not the author's conscious or unconscious intentions, is regarded as the proper focus of interpretation.)

The more economical the artistic style, the more interpretative work the audience has to do. Hence poetry, especially poetry of a highly minimalist type (such as haiku), is more tolerant of programmed production than prose is.

As an illustration of this point, consider the prose authored by a (very early) program designed to write detective stories. Although this system was a near-contemporary of the haiku-program, the aesthetic value of its productions was very much less. Instead of seventeen-word snippets, this program generated texts of over 2,000 words. Here are two extracts from one of them:[7]

The day was Monday. The pleasant weather was sunny. Lady Buxley was in a park. James ran into Lady Buxley. James talked with Lady Buxley. Lady Buxley flirted with James. James invited Lady Buxley. James liked Lady Buxley. Lady Buxley liked James. Lady Buxley was with James in a hotel. Lady Buxley was near James. James caressed Lady Buxley with passion. James was Lady Buxley's lover. Marion following them saw the affair. Marion was jealous.

> The day was Tuesday. The weather was rainy. Marion was in the park. Dr Bartholomew Hume ran into Marion. Hume talked with Marion. Marion flirted with Hume. Hume invited Marion. Dr Hume liked Marion. Marion liked Dr Bartholomew Hume. Marion was with Dr Bartholomew Hume in a hotel. Marion was near Hume. Dr Hume caressed Marion with passion. Hume was Marion's lover. Lady Jane following them saw the affair. Jane blackmailed Marion. Marion was impoverished. Jane was rich. Marion phoned Jane in the morning. Marion invited Jane to go to a theatre. Jane agreed. Jane got dressed for the evening. They met them in the theatre. Jane introduced Lord Edward during an intermission to Marion.

No literary prizes there! The reader's mind can barely manage to boggle, so deadening is the effect of such passages.

But what, exactly, is wrong with them? What would a more successful story-writing system need to be able to do which this one cannot? In other words, what computations in the minds of human authors would need to be matched in a convincing prose-writing program?

The difficulty of programming literary creativity (minimalist poetry excepted) is mostly due to three things: the complexity of human motivation, the need for background (commonsense) knowledge, and the complexity of natural language. Let us discuss these, in turn.

Human actions, motives and emotions – the usual concern of literature (and of everyday conversation and gossip) are even less easy to define than what an acrobat looks like. If Cohen's programs know less than we do about human bodies (being unaware that some unfortunates have only one arm), 'literary' programs know still less about what makes people tick. Moreover, people's moods and motives change continually, and storytellers are constrained to relate only changes that are psychologically plausible.

A major failing of the stories produced by the computerized Agatha Christie mentioned above is the shallowness of the motivations involved. For the program knows almost nothing about motivational structure.

The plots, such as they are, depend on a few simple constraints. These rule, for instance, that only couples who have previously flirted may be involved in a lover's tryst; trysts may take place once in the afternoon, in which case they may be observed, and/or once at night after everyone has gone to bed (matinal lovemaking is *unthinkable*). The opening lines of the story give thumbnail sketches of the characters:

Wonderful smart Lady Buxley was rich. Ugly oversexed Lady Buxley was single. John was Lady Buxley's nephew. Impoverished irritable John was evil. Handsome oversexed John Buxley was single. John hated Edward. John Buxley hated Dr Bartholomew Hume. Brilliant brave Hume was evil. Hume was oversexed. Handsome Dr Bartholomew Hume was single. Kind easy-going Edward was rich. Oversexed Lord Edward was ugly. Lord Edward was married to Lady Jane. Edward liked Lady Jane. Edward was not jealous. Lord Edward disliked John. Pretty jealous Jane liked Lord Edward.

These characterizations, chosen virtually at random, are used as constraints on what can follow. A mention of jealousy (or, later, flirtation) makes certain subsequent actions possible.

So far, so good. But the coherence of the story-line is minimal. Hatred, for instance, does not always have any result. And because this program (like abstract-AARON) is incapable of considering the text as a whole, as opposed to considering specific incidents in sequence, it cannot go back to prune 'redundant' hatred. That is, it cannot recognize loose ends as aesthetically unsatisfying, as human authors can.

Above all, the motivational structure of these stories is ludicrously simple. For example, the little drama concerning Lady Buxley and her lover James is not used to further (or intelligently to conceal) the main plot. Marion's jealousy has no outcome. Even when similar incidents have an outcome (such as 'Lady Jane yelled at Lord Edward'), this has nothing whatever to do with the performance, concealment or detection of the murder. In the story I have been quoting, the amorous James ends up being poisoned by his poor relation, the butler, who hopes to inherit his money. The notion that Marion might cooperate with the butler for reasons of her own is not one that the program's semantics can handle.

Similarly, the identification of the murderer comes as a statement rather than a discovery, there being no step-by-step detection – still less any deliberately planted false clues. Rather, the program waits until the story is long enough to guarantee that some of the characters will have reason to be at loggerheads, looks for such a pair, kills off one of them, and declares the other to be the murderer. Granted, the police and houseguests are described as 'looking for clues', and one character is announced to have found one; but there is no genuine or developed detection involved. Nor are any false clues planted, nor real clues slipped in unobtrusively or deliberately made ambiguous or misleading by the local context.

In short, the detective-novelist has no knowledge of the *general structure* of human motivation. It merely uses a few specific facts (that flirtation precedes lovemaking, for instance) at various points in the text. Consequently, it is not capable of producing genuine stories – not even at a childish level. As for the intricate story-line of Emily Brontë's *Wuthering Heights*, which involves complex interactions between characters of two generations (and the viewpoints of two different narrators), these are utterly out of its reach.

Since the detective-novelist program was written, there has been further AI-research on how to model the psychological phenomena that lie at the heart of most stories. Various aspects of motivation and behaviour can now be represented, at least in crude outline, by computational concepts such as *scripts*, *what-ifs*, *plans*, *MOPs* (memory organization packets), *TOPs* (thematic organization points) and *TAUs* (thematic abstraction units).

Scripts, what-ifs and plans were mentioned in Chapter 5. A script represents a type of social behaviour, defining complementary roles and sometimes specifying common variations (what-ifs). Scripts of various sorts (involving waiters and customers, doctors and nurses, cowboys and Indians . . .) help to shape virtually all stories.

A plan is a hierarchical structure of goals and sub-goals, constructed by means–end analysis. It may include contingency plans (what-ifs) anticipating certain obstacles. And it may represent some sub-goals schematically, leaving the precise details to be worked out at the time of execution. The intentions of the characters in a story are comparable to plans, and the author can add tension to the story by putting various obstacles in the way. (Frederick Forsyth's thriller *The Day of the Jackal* focuses not on 'who done it?' but on 'how will he manage to do it?' The villain's every sub-goal is achieved, largely because of ingenious forward-planning, until at the very last minute an unexpected – but highly plausible – event defeats him.)

A MOP is a high-level concept denoting the central features (details omitted) of a large number of episodes or scripts unified by a common theme. One example is 'requesting service from people whose profession is to provide that service'.

TOPs, too, are high-level schemata that organize memories and generate predictions about events unified by a common goal-related theme (such as 'unrequited love' and 'revenge against teachers'). But, unlike MOPs, they store detailed representations of the episodes concerned, rather than their thematic structure alone. (Psychologists make a similar distinction between 'semantic' and 'episodic' memory in the human mind.)

TAUs are abstract patterns of planning and plan-adjustment, each of which covers multiple instances (so can be used to remind the system of another story, superficially different but basically similar). Examples include: 'a stitch in time saves nine', 'too many cooks spoil the broth', 'many hands make light work', 'red-handed', 'hidden blessing', 'hypocrisy', and 'incompetent agent'.

The specific aspects of planning which are used to define TAUs, and recognize them within stories, are: enablement conditions, cost and efficacy, risk, coordination, availability, legitimacy, affect, skill, vulnerability and liability. Clearly, then, no program – and no human being either – can represent these high-level motivational concepts unless it can analyse the abstract structure of plans to this degree of detail.

These computational concepts have been implemented in a number of programs designed to interpret stories. The most impressive example to date is a program called BORIS, which also includes representations of facts about adultery and divorce, and the emotional and legal tangles they may involve.[8] BORIS uses its information about interpersonal phenomena such as anger, jealousy and gratitude to make sense of the episodes mentioned in the text.

(You may object to my saying that BORIS 'makes sense of' stories, or that it 'knows' about anger and jealousy. For BORIS does not *really* understand the stories it processes. I use these words, nevertheless, for two reasons. First, it is so much simpler than saying 'The program associates the word "jealousy" with "revenge", and this word triggers the construction of plans involving other words such as "lawyer".' Second, our prime interest is in what BORIS can teach us about human minds, which *can* make sense of these words. A program may embody psychological hypotheses about how concepts are used by people, without understanding those concepts itself.)

BORIS can give sensible answers to questions about a story in which a careless waitress, spilling soup on Paul's clothes, leads to his discovery (with a witness) of his wife *in flagrante delicto* in the conjugal bed. (Notice that BORIS, like you, needs to know some mundane background-facts in order to be able to do this: that clothes are usually kept in bedrooms, and that to get to one's house from a restaurant one may need to drive, or be driven, there.)

On reading the sentence 'Paul wanted the divorce, but he didn't want to see Mary walk off with everything he had,' BORIS can interpret 'walk off with' as meaning possession rather than perambulation. Also, it can see Paul's distaste for this prospect as a natural reaction to his discovery of his wife's infidelity.

Moreover, BORIS's knowledge of the origin and psychological

functions of various emotions enables it to assume that Paul's feelings will lead him to adopt certain strategies rather than others. In the story cited, Paul phones his friend Robert, a lawyer, to ask for advice. BORIS guesses the topic of the phone-call without its being actually stated, for it knows that help may properly be requested and willingly granted because of a past favour. (It does not assume that help *must* be forthcoming, for it knows that ingratitude is possible.)

If the story were to depict Paul as following some strategy different from any which BORIS had expected, a program similar to BORIS could not only be surprised, but could even criticize the episode as psychologically implausible. To be sure, we might reject its criticism as lacking in insight, for a program's sense of what is psychologically plausible can be no better than the psychological theory represented in it. But the same applies to human critics. One aspect of the novelist's creativity, for instance, is to enable readers to extend their sense of what is 'psychologically plausible' by exploring the conceptual space of human motivation more thoroughly than they had done before.

In short, BORIS suggests a positive answer to the third Lovelace-question – whether a computer could ever recognize (literary) creativity. The second Lovelace-question, however, demands more than BORIS can offer. If a program is to appear to be a creative *littérateur*, it must actually write stories, too.

Current story-writing programs can do better than the automatic detective-novelist, whose 2,000-word murder-mysteries were shapeless and rambling. The reason is that they involve better models of the psychological processes involved in story-writing.

These processes, of course, are the object of the exercise. The mystery-writer was intended as a preliminary exploration of the psychological issues involved, not as a way for the programmer to get rich by emulating Agatha Christie. It will be many years before AI-workers can sensibly aim to produce computer-generated stories that we might value for their own sake. Meanwhile, we should treat story-writing programs as aids to our understanding of the human mind, not as easy targets for our laughter or contempt.

Emily Brontë, then, is not in the picture. Occasionally, however, today's computers can do almost as well as Aesop.

A program called 'TALE-SPIN', for instance, produced this:[9]

George Ant was very thirsty. George wanted to get near some water. George walked from his patch of ground across the meadow through the valley to a river bank. George fell into the water. George wanted to get near the valley. George couldn't get near the valley. George wanted to get near the meadow. George couldn't get near the meadow. Wilma Bird wanted George to get near the meadow. Wilma wanted to get near George. Wilma grabbed George with her claw. Wilma took George from the river through the valley to the meadow. George was devoted to Wilma. George owed everything to Wilma. Wilma let go of George. George fell to the meadow. THE END.

A later version of TALE-SPIN, asked to write a story with the moral 'Never trust flatterers', came up with this:

The Fox and the Crow
Once upon a time, there was a dishonest fox named Henry who lived in a cave, and a vain and trusting crow named Joe who lived in an elm-tree. Joe had gotten a piece of cheese and was holding it in his mouth. One day, Henry walked from his cave, across the meadow to the elm-tree. He saw Joe Crow and the cheese and became hungry. He decided that he might get the cheese if Joe Crow spoke, so he told Joe that he liked his singing very much and wanted to hear him sing. Joe was very pleased with Henry and began to sing. The cheese fell out of his mouth, down to the ground. Henry picked up the cheese and told Joe Crow that he was stupid. Joe was angry, and didn't trust Henry any more. Henry returned to his cave. THE END.

Exciting, these little tales are not. But they have a clear structure and a satisfactory end. The characters have goals, and can set up sub-goals to achieve them. They can cooperate in each other's plans, and trick each other so as to get what they want. They can recognize obstacles, and sometimes overcome them. They can ask, inform, reason, bargain, persuade and threaten. They can even adjust their personal relationships according to the treatment they get, rewarding rescue with loyalty or deception with mistrust. And there are no loose ends, left dangling to frustrate us.

The reason is that TALE-SPIN can construct hierarchical plans, ascribing them to the individual characters according to the sorts of motivation (food-preferences, for example) one would expect them to have. It can think up cooperative and competitive episodes, since it can give one character a role (either helpful or obstructive) in another's plan. These roles need not be allocated randomly, but can depend on background interpersonal relations (such as competition, dominance and familiarity). And it can represent different sorts of communication between the characters (such as asking, or bargaining), which constrain what follows in different ways.

A story-writer equipped not only to do planning, but also to juggle with scripts, what-ifs, MOPs, TOPs and TAUs, could come up with better stories still.

Ideally, it would have more extensive knowledge about motivation than either TALE-SPIN or BORIS has: not only anger, jealousy, gratitude and friendship but (for example) shame, embarrassment, ambition and betrayal. To design such a program would be no small feat. Every psychological concept involved in the plots of its stories, whether explicitly named in the text or not, would need to be defined.

Consider betrayal, for instance, a concept that figures in many stories – from the court of the Moor of Venice to the Garden of Gethsemane. The theoretical psychologist must define 'betrayal' precisely, and also outline how the belief that one has been betrayed, or the memory of one's own betraying, can constrain a person's actions and emotions: Othello's jealous rage, or Judas' despairing suicide. How might this be done? And could it be done in a way that generalizes to other psychological categories?

A social psychologist (whose computational work influenced AI-research on scripts, MOPs and TAUs) has defined a number of interpersonal *themes* – of which betrayal is one – in terms of the interrelated plans of two people.[10] Each person, or *actor*, is thought of as having one plan: a means–end series of goals, with the logical possibility of obstruction or facilitation for each sub-goal. And each actor's relation to the other's plan has three logically independent dimensions: role, attitude and facilitative ability.

Every *role*, in turn, is defined in terms of three aspects. One actor can act as the other's *agent* (in respect of the whole plan or only certain parts of it), either temporarily or constantly; one actor may be *involved* in the other's goal (if the second actor plans to change, or to maintain, the current situation of the first); and one actor may be an *interested party* in the other's plan (if the latter's success would influence the former's opportunities to achieve his own goals).

An *attitude* is defined by the extent to which one actor approves or disapproves of the other's plan (in whole or in part), and of his own role in it. And *facilitative ability* is defined in terms of one actor's potential for helping or hindering the other's plan (in whole or in part).

Clearly, neither roles, attitudes nor facilitative ability need be reciprocal. Thus Wilma Bird was able to help George Ant by picking him out of the river, but the ant could not have done the same thing for the bird – and in some stories might not even have wanted to.

Moreover, whenever these concepts can apply to a plan 'in whole or in part', the part or parts in question must be specified. It is because people can do this that, for example, they can approve the end (the final goal of a plan) without approving the means (all the preliminary sub-goals).

These abstract dimensions define a conceptual space containing distinct types of interpersonal relation (see Figure 7.10). Each cell in the matrix identifies a different possible structure of actor-plan mapping, comparable to familiar psychological phenomena such as betrayal, cooperation, dominance and so on. Some cells allow for two or more such comparisons, marking the non-reciprocal nature of the relation in that cell (victory and humiliation) or the strength of the attitude involved (devotion and appreciation).

Figure 7.10

Sentiments Toward Other / *Influence of Actors*	Neither Influences Other	One Influences Other	Both Influence Other
Some Positive, No Negative	Admiration	(T_1) Devotion (T_2) Appreciation	(T_3) Cooperation (T_4) Love
One Actor Negative	(T_5) Alienation (also, Freedom)	(T_6) Betrayal (T_7) Victory (also, Humiliation) (T_8) Dominance	(T_9) Rebellion
Both Actors Negative	(T_{10}) Mutual Antagonism	(T_{11}) Oppression (also, Law and Order)	(T_{12}) Conflict

Betrayal, in this system, is defined as follows: *Actor F, having apparently agreed to serve as E's agent for action A, is for some reason so negatively disposed toward that role that he undertakes instead to subvert the action, preventing E from attaining his purpose.* In other words, the portion of conceptual space where betrayal can be found is bounded by these psychological constraints.

Is this structured definition of betrayal a good one? And how can it help us understand how human authors create their stories?

The definition may seem to imply that betrayal always has unfortunate consequences for E – in which case, stories in which E triumphs despite the betrayal would be literally unthinkable. But since plans are defined in terms of goals (or intentions), rather than successful actions, one can perfectly well allow that an action intended as a betrayal might fail to sabotage E's purposes.

Moreover, the plan concerned is one which is *attributed* (correctly or otherwise) by F to E. So we can conceive of betrayals involving actions, like Judas' kiss, that are directed against purposes mistakenly attributed to E. Jesus not only knew that Judas would betray him, but accepted it as a step towards the ultimate sacrifice.

Many story-lines (in fiction as in life) depend on a failure to specify the action (A) precisely. If one person thinks they have agreed to one thing, while the other thinks they have agreed to do something else, the result may be not only confusion but hotly disputed accusations of betrayal. And a systematic possibility obviously exists for making excuses, by redefining the range of the action A in defending oneself against charges of disloyalty.

Since betrayal is a structurally asymmetric theme (falling into the same classification-cell as the victory–humiliation pair), one might expect there to be two ways of describing it. Indeed, if one examines examples where actor E claims to have been betrayed, one rarely finds that actor F describes the incident in the same terms. Yet there is no one theme which reliably acts as the reciprocal of betrayal. Rather, actor F thinks of his action differently according to circumstances.

For instance, what both Montagues and Capulets saw as a betrayal of the family, Romeo and Juliet saw in terms of a different theme altogether (love), whose demands take precedence over usual family loyalties so that the concept of betrayal is out of place; what Czechoslovakia (in 1938) saw as a shameful betrayal, England represented as excusable prudence made necessary by her lack of armaments; and what Hitler saw as capital treason, von Stauffenberg regarded as justifiable action following a change of heart that unilaterally nullified the former contract between himself and the Führer. Presumably, the reason for the lack of a single reciprocal theme is that betrayal is morally disapproved, so people are rarely prepared to admit to it.

One way of exploring the conceptual space of betrayal is to vary the importance (to one actor or the other) of the actions involved. We can understand *abandonment* and *letting down*, for example, as distinct species of betrayal by 'tweaking' the definition given above.

To accuse F of abandoning E is to say that he was acting initially as E's agent for action A (this action being crucial to E's welfare); that he has now deliberately stopped doing so; and that this amounts in effect, if not necessarily in intent, to the deliberate subversion of E's purposes – since E (by hypothesis) is helpless without F. In contrast, to say that F let E down implies neither the urgency of A nor the helplessness of E.

In short, whereas anyone can let down or be let down, only the strong can abandon and only the weak can be abandoned. This is why abandonment is a peculiarly nasty form of betrayal.

Human authors, and readers, tacitly rely on such facts about the psychological structure of betrayal in writing and interpreting stories about it. The same applies to the other interpersonal themes shown in Figure 7.10. No computational system could create stories with any motivational depth without being able to construct and explore conceptual spaces of *at least* this degree of complexity.

This is not to say that the psychological theory summarized in Figure 7.10 is an adequate base for a convincing author, whether a person or a program. (Can you think of some interpersonal concepts that cannot be squeezed into this matrix by any amount of dimension-tweaking?) Other computational accounts of motivation and emotion have been suggested, one of which even provides an extensive lexicon of emotion-words classified in terms of the theory.[11] At present, however, there is no scientific theory providing a clearly defined place for all the psychological phenomena discussed in novels, drama and gossip.

It is hardly surprising, then, that current story-writing programs barely manage to match the psychological structures of infants' storybooks. Even the plot of a Barbara Cartland novel is a formidable challenge, and computer-generated stories involving interestingly complex motivation are not yet in sight. By the same token, current psychology cannot identify (even in outline) all the mental processes by which human authors produce their work.

The second obstacle to literary creativity in computers is the need for extensive background knowledge. Not every human author is like Frederick Forsyth, who spends many months doing detailed research in preparation for each new book. But all rely on a vast store of common knowledge shared, and taken for granted, by their readers.

All story-writing programs are provided with background knowledge of some sort. The detective-novelist knows a little about the relation between flirtation and trysts, and TALE-SPIN knows something about moving through space – hence 'Henry walked from his cave, across the meadow to the elm-tree.' (Compare Cohen's acrobat-AARON, which knows something about the changing shape of the muscles of the upper arm.) But they often fail to produce a coherent narrative because they lack elementary world-knowledge and common sense.

Frequently, they fail to make an inference that is so obvious to people that the programmer did not think of including any procedure to make it possible. Consider this 'mis-spun' tale from TALE-SPIN, for example:

Henry Ant was thirsty. He walked over to the river bank where his good friend Bill Bird was sitting. Henry slipped and fell in the river. He was unable to call for help. He drowned. THE END.

This was not the denouement intended by the programmer, who had expected that Bill would save Henry (in other words, that Bill would cooperate with Henry in solving Henry's problem). But he had not (yet) included any rule which enabled one character to *notice* what a nearby character was doing. Because of the rule that *being in water prevents speech*, Henry was unable to ask Bill to save him, or even to tell him that he had fallen into the river. So the unfortunate ant went, unnoticed, to a watery grave.

Many other mis-spun tales could be cited. Lack of elementary world-knowledge, on the part of at least one of the procedures within the program, can lead to this sort of thing:

One day Joe Bear was hungry. He asked his friend Irving Bird where some honey was. Irving told him there was a beehive in the oak tree. Joe threatened to hit Irving if he didn't tell him where some honey was. . . .

Sometimes, a program's lack of common sense can lead to a story containing an infinite loop:

Joe Bear was hungry. He asked Irving Bird where some honey was. Irving refused to tell him, so Joe offered to bring him a worm if he'd tell him where some honey was. Irving agreed. But Joe didn't know where any worms were, so he asked Irving, who refused to say. So Joe offered to bring him a worm if he'd tell him where a worm was. Irving agreed. But Joe didn't know where any worms were, so he asked Irving, who refused to say. So Joe offered to bring him a worm if he'd tell him where a worm was. . . .

The problem, here, was the program's sketchy understanding of goal-structure. (What story-writing heuristic – dealing with the attribution of goals to characters – do you think was added to prevent this sort of absurdity?)

Human authors usually take care of such matters without even consciously thinking about them. Occasionally, they slip up: some novelists unintentionally 'jump' a character from springtime in Paris in one chapter to the previous winter on the Adriatic in the next. This is why publishers' editors routinely check manuscripts for continuity of space, time, apparel and even proper names.

A story-interpreting program can therefore usually rely on the (human) author's having produced something sensible. A story-generating program, by contrast, has to do everything for itself. No wonder, then, if narrative disasters – like the tragedy of the drowning ant – crop up on every page.

The third reason for the difficulty of modelling literary creativity in computer programs is the complexity and subtlety of natural language: grammar, vocabulary and meaning.

Anyone who thinks that grammatical skill has nothing to do with creativity should remember the programmed detective-stories quoted above. Not even the most unimaginative hack authors write such boringly unstructured sentences. Their aesthetic value is nil.

The prose actually produced by TALE-SPIN is no better, for the program's linguistic module (aptly named 'MUMBLE') cannot cope with compound sentences, or with pronouns. The perfectly acceptable sentence 'But Joe didn't know where any worms were, so he asked Irving, who refused to say' is a tidied-up version produced by the programmer. MUMBLE's equivalent was something like this: 'Joe didn't know where any worms were. Joe asked Irving where the worms were. Irving refused to tell Joe where the worms were.'

In fairness, the programmers concerned were not interested in the grammatical capabilities of their programs, only in generating plots. But the fact remains that a convincing computer-author would need to be able to write grammatically complex prose.

Much as programs for interpreting stories (or melodies) are, in general, more successful than programs for writing them, so most language-using programs can only parse – not produce – elegant syntax. A few, however, can generate sentences with highly complex grammatical structure. For instance, a program playing noughts and crosses came up with the following passage to describe the game indicated in Figure 7.11:

> I started the game by taking the middle of an edge, and you took an end of the opposite one. I threatened you by taking the square opposite the one I had just taken, but you blocked my line and threatened me. However, I blocked your diagonal and threatened you. If you had blocked my edge, you would have forked me, but you took the middle of the one opposite the corner I had just taken and adjacent to mine and so I won by completing the edge.[12]

Figure 7.11

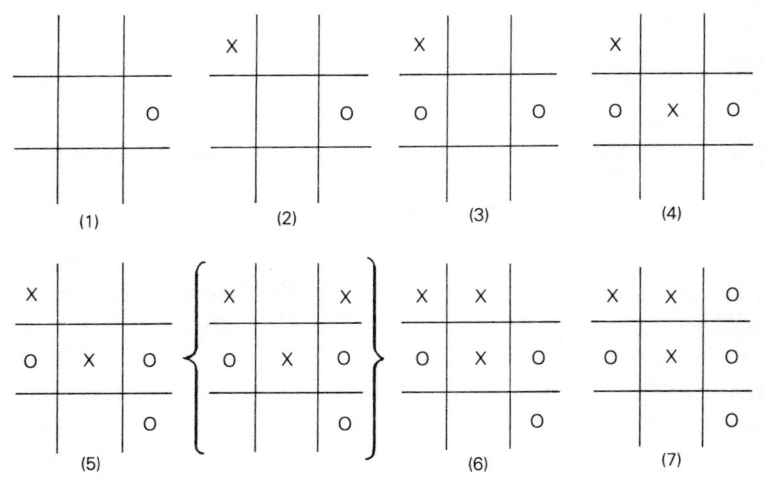

The grammatical subtleties here are considerable. The boring repetition of nouns is avoided by using words like *one* and *mine*, and a string of 'buts' is avoided by using *however* instead. The relative times of distinct events are shown by tensed verbs like *had taken*. And there are many compound sentences, made with the help of words such as *and*, *but*, *however*, *if* and *so*.

Even more significant, the syntax is semantically appropriate: the main and dependent clauses itemize the more and less important points, respectively.

For instance, the second sentence would have been much less apt if the order of the last two ideas had been reversed: I threatened you by taking the square opposite the one I had just taken, *but you threatened me and blocked my line*. The order actually chosen by the program reflects, in a natural fashion, the structure of attack, defence and counter-attack informing this game.

Similarly, it would have been less appropriate to express the first half of the same sentence like this: I took the square opposite the one I had just taken and so threatened you. For the syntax of subordinate and subordinating clauses actually chosen by the program corresponds to the strategic importance of the ideas involved: *that* I threatened you is more important than *how* I did so, and so should be the main focus of the sentence.

This rule is apparently broken in the next (third) sentence of the game-description: 'However, I blocked your diagonal and threatened you.' But this is as it should be, because the blocking of the diagonal was the necessary defensive response to the previous threat from the opponent, whereas the fact that it also constituted a new threat *to* the opponent was a fortunate side-effect. (Similar remarks apply to the second half of the second sentence.)

The program's choice of 'However' (in the third sentence) was guided by the rule that two consecutive *buts* within a single sentence should be avoided, because they may be confusing to the reader. Accordingly, *however* was used at the start of a new sentence, signalling that the blocking of the opponent's diagonal depended (strategically) on the previous threat posed by the opponent. That is, the choice of two sentences rather than one was not a matter of mere stylistic elegance (like preferring the phrase *a vain and trusting crow* to *a trusting and vain crow*), but was aimed at helping the reader's interpretation in a specific way.

Admittedly, this program could not write a page-long sentence without getting lost. But many human writers cannot do so either. Indeed, we marvel at Proust's ability to sustain his syntax over long periods.

Admittedly, too, it is much easier to say what is 'important' in noughts and crosses than in human motivation. TALE-SPIN's knowledge of planning would be enough to enable a grammatically adept computer to put the *but* and *so* into the sentence about Irving's refusal to say where the worms were. A BORIS-like writing-program could handle its prepositions even better. But to emulate Proust's use of syntax, a program would need his extensive knowledge of psychology as well as his mastery of grammar.

In short, grammar is essential to literary creativity. It prevents us from being bored by verbal repetition. Even more to the point, it helps in subtle ways to guide us through the conceptual space presented to us by the author.

Vocabulary and meaning are widely recognized as crucial to literature. The exceptional range of Shakespeare's vocabulary, for instance, is often remarked. Unlike the computerized detective-novelist, he could report a lovers' tryst in many ways besides 'X caressed Y with passion', or 'X was with Y in a hotel'. So could you, of course.

It takes each of us many years of listening and reading to stock our word-store, and literary experts who can be relied on to find *le mot juste* devote even more time to this project. A comparable computer-program would need tens of thousands of words.

Even more important than a wide vocabulary, an author needs sensitivity to the underlying meanings that enable us to link one word with another in creative ways. Readers need this sensitivity too, if they are to recreate the author's ideas in their own minds. Think of Macbeth's description of sleep, for instance:

> Sleep that knits up the ravelled sleeve of care,
> The death of each day's life, sore labour's bath,
> Balm of hurt minds, great nature's second course,
> Chief nourisher in life's feast.

This passage works because Shakespeare's readers, like him, know about such things as knitting, night and day, and the soothing effects of a hot bath.

Or consider the literary conceit in Plato's *Theaetetus*, where Socrates describes himself as 'a midwife of ideas'. Socrates (he explains to Theaetetus) is too old to have new philosophical ideas, but he can help his pupils to have them. He can ease Theaetetus' labour-pains (his obsession with a seemingly insoluble philosophical problem). He can encourage the birth of true ideas, and cause the false to miscarry. He can even match-make successfully, introducing non-pregnant (non-puzzled) youngsters to wise adults who will start them thinking. His skill, he says, is greater than the real midwife's, since distinguishing truth from nonsense in philosophy is more difficult than telling whether a newborn baby is viable.

Sleep and knitting, philosophy and midwifery. Could computers even appreciate such verbal fancies, never mind come up with them?

Well, they already have. A computer model of analogical thinking called ACME ('M' is for Mapping) has interpreted Socrates' remark appropriately, showing *just how* a midwife's role in aiding the birth of a new baby resembles a philosopher's elicitation of ideas in his pupil's mind – and how it does not. (The match-making comparison is not picked up; but this is no longer part of the meaning of midwifery, so a philosophy-student today would not think of it either.)[13]

So far as I know, this program has not been tried out on Macbeth's speech about sleep. But I'd be prepared to bet that it could make something of it. For ACME uses highly abstract procedures for recognizing (and assessing the strength of) analogies *in general*. Moreover, it has an unusually large vocabulary, whose underlying meanings are stored in a relatively rich semantic network.

Its memory is like a dictionary and thesaurus combined, whose concepts are analysed in much greater detail than is usual. This network, an independently developed (not tailor-made) system called

WordNet, takes account of detailed psychological studies of language and human memory. It already contains over thirty thousand word-entries, to which more can be added at will (much as one might explain the unfamiliar word 'balm' to a schoolchild reading *Macbeth*).

WordNet is a connectionist system in which each unit represents a concept. (Using the distinction introduced in Chapter 6, this is a 'localist' memory, not a 'distributed' one.) The links between units code abstract semantic features such as *superordinate*, *subordinate*, *part*, *synonym* and *antonym*. These abstract features appear to be coded also in the human mind, which uses them for mental explorations of various kinds. For example, you may remember our discussion, in Chapter 4, of the heuristic 'consider the negative': given a semantic memory in which antonyms are routinely stored, the 'negatives' of concepts will be readily accessible.

In this way, WordNet provides a rich penumbra of meaning for every conceptual item. The concept of animal, for instance, is directly linked to many other concepts – including organism, living-thing, prey, person, child, mammal, primate, reptile, fish, bird, insect, vertebrate, game, voice, tooth, claw, beast, creature, fauna, plant and flora. Indirectly, it is linked to many more.

Having a large vocabulary, and a rich store of meanings, is all very well. But it could prove an *embarras de richesse* for a mind, or a machine, wanting to compare one concept with another.

Consider Socrates' analogy, for instance. Let us assume – a gross oversimplification – that 'philosopher' and 'midwife' each have links to only five other nouns (such as 'idea', 'pupil', 'cradle', 'kettle', 'mother' and 'baby') and only ten adjectives (such as 'intellectual', 'medical', 'eliciting' and 'gentle'). Let us assume, too, that when comparing two concepts, nouns can be mapped only on to nouns and adjectives only on to adjectives. In that event, there would be over 400 million possible comparisons between 'philosopher' and 'midwife'.

Yet ACME, like Plato's readers, manages to understand the analogy and can explain it if required. How is this possible?

The answer lies in a computational technique explained in Chapter 6, namely, multiple constraint-satisfaction. A connectionist network trying to understand an analogy can consider many different constraints in parallel, and settle on the best available match even though this match may be somewhat flawed (and, by hypothesis, is never perfect).

The general constraints considered by ACME, when mapping one half of an analogy on to another, are of three types. The first is *structural consistency*: the program favours analogies where there is a one-to-one mapping of all the elements in the two halves. So if a philosopher is to be mapped to a midwife, then something-or-other should map on to a baby; and ideally, if sleep is mapped to a knitter, something should map to the knitting-needles. (Likewise, a necklace-game enabling one to do subtraction as well as addition would be a better analogue of arithmetic than the game defined at the outset of Chapter 4.)

The second analogical constraint is *semantic similarity*: ACME favours mappings between elements that have similar meanings. On this criterion, philosopher/midwife (both animate, human beings) is a better mapping than idea/baby – even though both the idea and the baby are new, and somewhat fragile.

The third constraint on understanding analogies is *pragmatic centrality*. The program (like us) prefers mappings that it judges to be important to the analogist, either because a correspondence between two specific elements has been stated to hold, or because an element is so central to its parent-structure that *some* mapping for it needs to be found. For instance, Shakespeare tells us – much to our surprise – that sleep is a knitter, and that sleep is a bath; so our interpretations must strive to preserve those surprising correspondences. Again, a baby is so central to the concept of a midwife that something must be mapped on to it, even if that something is non-human and non-animate, like an idea.

As these examples of human thinking show, pragmatic centrality can override the other two criteria. Never mind that sleep has no needles, or that ideas are inanimate: 'knitter' and 'baby' must appear somewhere in the mapping, if the analogy is to be accepted. Consequently, ACME is designed so as to give special emphasis to the pragmatic criterion.

Most analogy-programs ignore pragmatic centrality, focusing instead on structural and/or semantic similarity. Moreover, most insist on certain correspondences, and cannot find a satisfactory interpretation if correspondences conflict. ACME's three abstract constraints, together with its use of multiple constraint-satisfaction, makes it much more powerful as an analogy-recognizing mechanism.

The program's interpretation of analogies fits our description of creativity as exploration. In effect, ACME is exploring its own conceptual space, simultaneously using 'mental maps' of three different types to guide it. The three constraints (given the background associative memory) provide a generative system, a way of building specific similarity-structures that did not exist before.

Because ACME uses multiple constraint-satisfaction, it is able to set aside many normal considerations – so as to see a pupil's new philosophical idea as something like a baby, for instance. One might even say that it *transforms* the idea into a baby, but this transformation (unlike the transformation of benzene-strings into benzene-rings) is only temporary. It is, after all, only an analogy.

Programs for doing analogical mapping are relevant to the third Lovelace-question, since they enable computers to understand – and even to assess – surprising conceptual comparisons. They help answer the second Lovelace-question too, since someone with a new idea must evaluate it himself (as Kekulé had to do, with respect to his adventurous idea about the benzene molecule). But creativity also requires the *generation* of analogies. What of that?

Many convincing analogies have been P-created by a program called ARCS ('R' is for Retrieval), which is very closely related to ACME.[14] What ARCS does is to find an analogy for an idea that is suggested to it, much as a schoolchild might be asked to think of an analogy for 'winter'. The idea in question may be a concept, like winter, or a series of propositions, like the plot of a story.

For instance, given a plot-outline of *West Side Story*, and twenty-four synopses of Shakespeare's plays to choose from, ARCS picked out *Romeo and Juliet* as the closest. And, given a story about a person deciding that an unattainable goal was not desirable anyway, it retrieved 'The Fox and the Grapes' from one hundred of Aesop's fables; this is especially impressive in view of the strong similarity between any two of Aesop's tales (most of which, for instance, involve animals as characters).

Because its retrieval-procedures are highly abstract, ARCS can P-create analogies in many different domains – including scientific problems. Thus it solved the medical problem of how to destroy a tumour with X-rays without damaging the healthy tissue surrounding it, by picking the closest of five problems involving laser-beams, ultrasound and an army attacking a fortress. (This problem has been used for many years in experiments on human problem-solving; ARCS' performance is significantly similar to the experimental results in various ways.)

If analogies are to be intelligible (to the creator or to some third party), then the processes that generate them must exploit mapping-constraints and memory-structures similar to those used by the processes that evaluate them. In that sense, the creation and appreciation of analogies take place within the same conceptual space.

But much as a human reader has to do less work in understanding analogies than an author does in producing them, so the analogy-mapping program has less work to do than the analogy-retriever. ACME is presented with a specific analogy, and has to make sense of it. But ARCS has to find an analogy for a given concept, without any hints from outside as to which information in its memory (WordNet again) may be relevant. There are always many possible choices. (Open your dictionary and pick two concepts at random: very likely, they will have *some* degree of similarity.) Somehow, ARCS must pick the best one.

It does this in two steps. First, it finds a large set of potential analogues which are semantically similar to the input-concept, and uses multiple constraint-satisfaction to whittle these down to the few whose meaning is closest of all. Second, it considers these few in terms of all three general constraints: not just meaning, but structure and pragmatic importance too. This step, likewise, involves multiple constraint-satisfaction. In this way, ARCS finds the best match overall (even though other candidates may be better on a single criterion).

You probably feel that the program-generated analogies, and Socrates' analogy too, are less creative than the comparisons (in Macbeth's speech) of sleep with a knitter or a hot bath. – Why? How does this literary intuition fit with the definition of creativity introduced in Chapter 3? Can we say that the program's (and Socrates') novel ideas could have happened before, whereas Shakespeare's new ideas could not? Or must we allow that all these analogies are equally creative? After all, a philosopher *is not* (usually!) a midwife, and sleep *is not* a hot bath. Why do we regard one as more imaginative than the other?

The key, here, is the extent to which the semantically central features of one half of an analogy are matched in the other half. In the Shakespearean examples, a central feature of one analogue is not matched in the other, the analogy being carried only by the more peripheral features.

Thus to compare one animate, human being with another (midwife/philosopher) is less creative than to compare a mode of consciousness (sleep) with a human being (knitter) or an inanimate object (bath). To think of a philosopher or midwife – or a knitter – in the literal sense, one must think of them as animate. But the literal meaning of sleep does not allow us to think of it in that way. To do so, is to do something which *could not* be done according to the normal constraints on meaning.

'Poetic licence' enables poets to ignore familiar constraints of meaning (and of truth), at the cost of requiring the reader to do more interpretative work. Because its creators are interested in the philosophy of science (and have applied ARCS to some historically important scientific concepts), ARCS is designed to pick the closest analogy, given a certain pragmatic context. But the poet looks for more distant analogies, which force us to think about a concept in new ways. It is precisely because a knitter is not a close analogue to sleep that Shakespeare, like his readers, found the comparison interesting. If ARCS were instructed to reject the twenty closest analogies, it would come up with some surprising items too.

Moreover, ARCS is designed to highlight one analogy at the expense of others. If it found 'knitter', 'death' and 'bath' among the potential analogues for 'sleep', it would pick death and forget about the rest.

Shakespeare, by contrast, offers us a profusion of intelligible analogies. To compare sleep with human beings, death, hot baths, ointment and the main course of a meal (not to mention 'hurt minds' and 'life's feast') – all within a single sentence, expressed in a few lines of blank verse – is to awaken so many of the ideas latent in our minds that we experience a glorious explosion of newly recognized meanings.

Marvellous, this is. But magical, it is not. Shakespeare's rich store of vocabulary, meaning and everyday knowledge – even such mundane facts as that meals in England have several courses, of which the second is the most sustaining – was an essential source of the lines quoted above. Indeed, it presumably influenced the sequence in which these specific images arose in his mind.

For instance, he often used 'course' with reference to the sun's movement, and the phrase 'death of each day's life' may have triggered this sense of the word. As he wrote it, however, the food-related sense (which was current in his time) may have been excited also, leading in turn to the idea of 'life's feast'. The expression 'the course of nature' (also used in Elizabethan times) may have been involved too: once 'course' had been triggered, it could have been directly associated with 'nature'.

Shakespeare's familiarity with the constraints of English grammar and iambic pentameter were essential too. Somehow, the computational processes at work in his mind managed to integrate this rich and diverse knowledge with general procedures for analogical mapping. These procedures were certainly much more powerful, and much more subtle, than any current program. But they may have been broadly comparable.

Analogy is crucial not only to literature, but to the visual and performing arts, to science and mathematics, to much humour and to everyday chit-chat.

So ACME and ARCS help us to understand, in general terms, how it is possible (for instance) for a ballerina to portray a wooden doll and for the audience to interpret her dancing accordingly. The specific concepts involved are largely visual, like those used by the acrobat-drawing program. But as well as dealing with the shapes and positions of the limbs, they must discriminate between jerky and fluid movements. Without such distinctions, the ballet-audience would not be able to interpret the scene in *Coppelia* where the magician turns the doll into a real girl.

Likewise, these computational models help us to see what sort of mental processes are involved in understanding the relevance of the necklace-game. The child spontaneously *retrieved* the analogy of addition, and found its superordinate category, arithmetic (which includes subtraction as well). Probably, you did so too; and you may even have gone on to retrieve the still-higher category of number theory.

Having retrieved (or having been given) the mathematical analogy, the child – and you – then went on to *map* it against the necklace-game, pushing it as far as possible until its bounds were reached. In other words, the analogy was not 'idle'. It enabled you and the child to explore the conceptual space of the necklace-game in a disciplined, and fruitful, way.

Science uses many non-idle analogies. Kekulé's tail-biting snake, alias the benzene molecule, is one example. The Bohr solar-system model of the atom is another. You will be able to think of many more. Indeed, Koestler regarded 'the real achievement' in many scientific discoveries as 'seeing an analogy where no one saw one before'.

As for how this happens, Koestler commented (as we saw in Chapter 2):

> [In most truly original acts of discovery the analogy] was not 'hidden' anywhere; it was 'created' by the imagination. . . . 'Similarity' is not a thing offered on a plate [but] a relation established in the mind by a process of selective emphasis. . . . Even such a seemingly simple process as recognizing the similarity between two letters 'a' written by different hands, involves processes of abstraction and generalization in the nervous system which are largely unexplained. . . .[15]

How right he was! By a coincidence which he would have relished, ACME and ARCS were largely inspired by neural networks designed to recognize analogies – including typographic styles on the one hand and individual letters of the alphabet (in whatever typeface) on the other.[16] More important, the actively selective comparison-procedures employed in these programs support Koestler's insistence that similarity is not offered on a plate. You had to *think*, to see the mathematical implications of the necklace-game. So did Kekulé: first to see snakes as molecules, and then to map them on to each other so as to allow for valency.

Koestler's comments, then, were well taken. His appeal to 'the bisociation of matrices', vague though it was, pointed in the right direction. But we now have something which he did not: the beginnings of a rigorous explanation of how analogical thinking is possible.

Analogy is not the only sort of thinking that pervades arts and sciences alike. Induction does, too. Everyone makes generalizations on the basis of limited evidence. Notions like 'Baroque music' or 'Gothic architecture' classify a host of individual examples (including some we have never actually seen), while concepts like 'acid' or 'specific heat' enable us also to predict and to explain. The processes involved in these kinds of thinking are discussed in the next chapter.

8

Computer-Scientists

Let us begin with a story:

Once upon a time, there were two neighbouring soybean-plantations. Five years in succession, the plants showed symptoms of disease (different each year).

One plantation-owner was rich, and could afford to consult the person who knew more about soybean-diseases than anyone else in the world. The other plantation-owner was poor, and had to seek advice elsewhere. Whereas the rich man could send his private jet to fetch the world-expert to his plantation, the poor man had to use the telephone or the postal service. And while one could pay for microscopic examinations, the other could not.

Yet, each year, the poor owner's plants responded to treatment while the rich owner's plants sickened and died. As the poor owner flourished, his wealthy neighbour struggled. By the sixth harvest, their fortunes were reversed.

This tale is a fiction. But it is not a fairy-story. The poor owner had no genie or fairy-godmother looking after his interests. Rather, he had sought advice from a computer program, paying much less for this service than the human expert's consultation-fee. Nor is it science-fiction. It is not premised on some daring imaginative leap, such as a time-machine or anti-gravity device. For a computer program already exists which gives near-perfect diagnoses of most soybean diseases, and which surpasses the 'textbook' method defined by the world-expert.[1]

Agricultural experts can diagnose soybean-diseases 'intuitively'. While rare conditions may require microscopic study, the nineteen common diseases can be identified by means of easily observable features such as the plant-parts affected and the abnormalities they display. Leaf-spots, for instance, may be large or small, with or without haloes or watersoaked margins. In general, there is no simple (one-to-one) relation between symptom and disease: each disease exhibits a complex pattern of symptoms, which human specialists learn to recognize.

The specialists' advice is made available – at a price – to individual farmers, who normally describe their problem by phone-call or letter. But the farmer may not know just what signs to look for. And the expert (who could solve the problem instantly were he to visit the farmer's fields) is not always immediately able to ask the specific questions that would solve the problem. Consequently, AI-workers suggested some years ago that a specialist AI-program (an 'expert system') might be helpful.[2]

They asked a prominent authority on soybean-disease to tell them what evidence he used in making his diagnoses. During many hours of discussion, he described his intuitive skill as explicitly as possible. The diagnostic methods, or heuristics, he provided were then embodied in a computer program. They were represented as a set of IF–THEN rules, linking evidence to plausible conclusions: IF there are large leaf-spots, THEN it may be one of *these* diseases, but not one of *those*; IF the leaf-spots have yellow haloes, THEN *such-and-such* diseases are ruled out, but *these others* are possible.

To use the program, a farmer's problem would be described by a questionnaire itemizing the (thirty-five) relevant features. This data would then be input to the program, which used its IF–THEN rules to find the diagnosis (an example is shown in Figure 8.1). Tested on a set of 376 cases, the program got 83 per cent of its diagnoses right.

You are probably not impressed. Quite apart from the 17 per cent error-rate, every rule used by the program was specifically provided by the human expert. It had to be told everything, being able to learn nothing. If any program is subject to Lady Lovelace's criticism (that it can do 'only what we order it to perform'), this one is.

People are different. Given time, a farmer can learn to recognize soybean-diseases by being shown a variety of examples and counter-examples. His teacher need provide no explicit rules (if he tries to do so, some of them may even be mutually contradictory). Instead, he points out the features – spots and so forth – relevant to the diagnosis in each individual case. Given this sort of help, people can P-create their own concepts.

Previously, the farmer had probably never heard of 'frog eye leaf spot'. Even if he had, he could not recognize it. He could not locate it (even implicitly) on any conceptual map. Now, he can. He does so by using lower-level concepts such as 'watersoaked margins', some of which may be newly learnt too. Often, he learns not just what pathways to take when navigating in the new space, but in what order. For instance, he may come to look for haloes before worrying about watersoaked margins, because – with respect to soybean-disease – haloes are more informative.

Figure 8.1

Environmental descriptors
- Time of occurrence = July
- Plant stand = normal
- Precipitation = above normal
- Temperature = normal
- Occurrence of hail = no
- Number years crop repeated = 4
- Damaged area = whole fields

Plant global descriptors
- Severity = potentially severe
- Seed treatment = none
- Seed germination = less than 80%
- Plant height = normal

Plant local descriptors
- Condition of leaves = abnormal
  - Leafspots–halos = without yellow halos
  - Leafspots–margin = without watersoaked margin
  - Leafspot size = greater than $\frac{1}{8}''$
  - Leaf shredding or shot holding = present
  - Leaf malformation = absent
  - Leaf mildew growth = absent
- Condition of stem = abnormal
  - Presence of lodging = no
  - Stem cankers = above the second node
  - Canker lesion color = brown
  - Fruiting bodies on stem = present
  - External decay = absent
  - Mycelium on stem = absent
  - Internal discoloration of stem = none
  - Sclerotia–internal or external = absent
- Conditions of fruits–pods = normal
  - Fruit spots = absent
- Condition of seed = normal
  - Mould growth = absent
  - Seed discoloration = absent
  - Seed size = normal
  - Seed shrivelling = absent
- Condition of roots = normal

Diagnosis:

Diaporthe stem canker( ) Charcoal rot( ) Rhizoctonia root rot( ) Phytophthora root rot( ) Brown stem root rot( ) Powdery mildew( ) Downy mildew( ) Brown spot(×) Bacterial blight( ) Bacterial pustule( ) Purpose seed stain( ) Anthracnose( ) Phyllosticta leaf spot( ) Alternaria leaf spot( ) Frog eye leaf spot( )

(Whether he realizes that he is thinking in this way is another matter. In general, one's ability to *describe* the structure of a conceptual space one inhabits is limited. Thus even though the soybean-specialist tried hard to express his expertise in explicit form, the resulting rules were successful in only 83 per cent of cases; clearly, much expertise remained tacit. As we saw in Chapter 4, the ability to reflect on one's mental processes originates in early childhood; by definition, it is a step behind those mental processes themselves.)

You may feel that the expert system should be more like the farmer, that an ability to learn new concepts by example is the least one would demand of a 'creative' program. Well, this demand has already been met.

Many *inductive* programs exist, which have come up with newly defined general rules on the basis of collections of particular instances. Examples focusing on famous cases of scientific H-creativity will be discussed soon. For the moment, let us stick with soybean-disease.

In an early instance of automated induction, 307 ailing soybean-plants were described by the questionnaire shown in Figure 8.1 and each one was diagnosed by a human specialist. The 307 description–diagnosis pairs were then input to a simple inductive program, which searched for regularities in the mass of data presented to it. After this training-experience, the program was tested by a different set of plant-descriptions (now, with no ready-made diagnoses attached).

Tried out on 376 cases, it gave the wrong diagnosis in only two of them. That is, its self-generated rules achieved almost 100 per cent accuracy. The inductive program's newly defined diagnostic rules were more efficient than the human specialist's 'textbook' methods (embodied in the soybean-program described above), which managed only 83 per cent success on the same test-set. The program-generated rules have therefore been put into an updated soybean-program, which is now in routine use as a diagnostic tool in the Illinois agricultural service.

Positive value, as we have seen, is a criterion of creativity. A program with a near-100 per cent success-rate in performing a socially useful task is not to be sneezed at. Nor is percentage-success the only relevant value in assessing new ideas: simplicity often matters, too. The inductive program produced a systematically constructed concept, not a rag-bag of individual rules. Indeed, its newly produced concept of soybean-disease was the *best possible* representation of the data input to it, as we shall see. (An essentially similar inductive program has defined new chess endgame rules that are more elegant, and more readily intelligible to human players, than those given in the chess-books.[3]) The criterion of elegance has apparently been met.

How was this achievement possible? Unlike the (original and updated) 'expert systems' described above, the inductive program knew nothing substantive about soybeans. It used a purely logical approach to find abstractly defined regularities in the data, irrespective of subject-matter. In brief, it examined the data to find features that were always (or sometimes, or never) associated with a given diagnosis, ensuring that all individual features and diagnoses were considered.

This approach, expressed as the 'ID3' algorithm, is employed in many inductive programs. (ID3 can also 'tidy up' a set of human-derived rules, like those used to write the 83-per-cent-successful soybean-program, by identifying any hidden inconsistencies.) Provided that the number of relevant properties is not so large as to make even a computer suffer from information-overload, ID3 is *guaranteed* to find the most efficient method of classification in a given domain. In other words, there is a mathematical proof that (given enough time and memory) the algorithm can, in principle, do this. But the set of specified properties must include all – though not necessarily only – the relevant ones. A farmer, by contrast, might notice that yellow haloes are relevant even if they are never specifically mentioned to him.

A learning program that uses this logical strategy can structure a 'classification-by-property' conceptual space in the most economical way, and can find the shortest pathway for locating examples within it. In the terminology introduced in Chapter 5, the program defines not only the relevant search-tree but also the most efficient tree-search. It learns to ask the right questions in the right order, so as to decide as quickly as possible (for instance) which of the nineteen soybean-diseases afflicts the plant in question.

This depends partly on the relative numbers of the different classes (diseases) in the example-set considered as a whole. Suppose, for instance, that having yellow haloes is a *sufficient* condition for a diagnosis of frog eye leaf spot. Even if it is just as easy to check for yellow haloes as for any other property (which in practice may not be true), it does not follow that the most efficient decision-procedure will start by looking for them. For that particular property may be possessed only by a tiny percentage of plants suffering from frog eye leaf spot.

Even if yellow haloes were a *necessary* condition of this disease, it might sometimes be sensible to consider other properties first. This would be so, for instance, if examples of frog eye leaf spot were comparatively rare on soybean-farms.

The ID3 algorithm can identify, and then exploit, statistical properties like these. Clearly, it must be shown a *representative* sample, in which the common diseases predominate and the rarer diseases are correspondingly few. Otherwise, the diagnoses it learns to make will be unreliable.

How does such a program compare with a human farmer learning to recognize plant-diseases? Like him, it can make diagnoses at the end of the learning process which it could not have made before. And like him, its reliability depends on its having experienced a representative sample of soybean-diseases.

But there are differences, too. Because it lacks the human's short-term memory limitations, the program can search very much larger example-sets than we can. It has no special difficulty in processing negatives or disjunctives. By contrast, people find it relatively difficult to use the information that a disease *does not* involve a certain symptom, or that a plant with that disease will show *either this symptom or that one.* And it ignores the fact that, in practice, some relevant properties are more difficult to identify, less salient, than others. For instance, it does not know that small spots are less easy for human beings to see than larger ones are. In short, the program P-creates its concepts in a somewhat inhuman way.

This does not mean, however, that the program is psychologically irrelevant. Its general approach, or logical strategy, is one which people can use too (consciously or unconsciously). Indeed, the inductive algorithm was initially suggested by psychological research on how people learn concepts.

It shows us that a theory of human concept-learning which (while also taking short-term memory into account) appealed to search-trees and computations like those represented in the program *could* explain a wide variety of human successes. And it offers a clearly defined theoretical structure within which psychologists can explore different 'weightings' of properties with varying perceptual salience. If yellow haloes are so obvious that they 'hit one in the face', whereas watersoaked margins are not, these facts about ease of (perceptual) processing can be represented in the computational theory concerned.

Moreover, the statistical insights that inform such inductive programs could be represented in a neural network. This is a special case of a general point made (with reference to musical interpretation) in Chapter 5: heuristics initially defined in 'inhuman' logical-sequential programs can be embodied in parallel-processing systems that are more tolerant of noise. Such a system might, for instance, diagnose frog eye leaf spot successfully even in cases where the (usually definitive) yellow haloes were missing.

Nor can one say that the program is irrelevant because it deals with what is, to most people, a thoroughly boring subject – namely, soybean-diseases. For soybean-diseases, like chess endgames, are just examples. An art-historian could have provided ways of recognizing a Michelangelo sculpture or an Impressionist painting. A literary critic could have defined epic poetry. Or a musicologist, sonata-form. Like analogy-programs (such as ARCS and ACME), the ID3 algorithm is in principle relevant to concepts in *any* domain.

Some readers will doubtless object that candidates for creativity should produce H-creative concepts, not merely H-creative rules for defining pre-existing concepts. 'Agricultural experts', they will say, 'already knew about frog eye leaf spot. Moreover, the inductive program was given this class-concept for free (though admittedly it produced a better definition of it than any human had). The person who identified this disease in the first place was the *really* creative conceptualizer. No computer program could come up with a brand new concept, unknown to any human being.'

Unfortunately for our imaginary objector, some AI-systems have already done so. Indeed, the ID3 algorithm itself has discovered useful regularities of a fairly complex kind, previously unknown to human experts. We noted above that its input-data (concerning the classification of soybean diseases, for instance) usually mentions only properties known to be relevant. But, provided that it can tell whether an example falls into a certain class, ID3 can assess a property's relevance itself.

For instance, an ID3-program for playing chess (wherein examples of 'a win' and 'a loss' can be very easily recognized) used seemingly irrelevant input board-descriptions in finding previously unknown winning strategies for chess endgames. Or rather, it used board-descriptions whose specific relevance – if any – was unknown to the human chess-master who provided the input to the program.

The chess-master suspected, for example, that the white king's being on an edge is relevant in endgames involving king and pawn against king and rook. But he did not know how. Moreover, no chess-master anywhere had managed to define a winning strategy for this particular endgame, whose complexity (potential search-space) is too great for the unaided human mind.

With the help of ID3, however, the task has been partly achieved. Provided that the pawn starts on – or can be manoeuvred on to – a certain square, this endgame can be won by a strategy consisting of nine rules (easily intelligible to chess-experts), ordered by means of a search-tree. In sum, the computer-generated concept of *a winning strategy for this endgame* is H-novel, successful, useful (to some people) and elegant. A chess-master who had come up with it would have been widely praised.

To be sure, a human chess-player would have been less 'logical'. He could not have been sure that all the possibilities had been allowed for. But to allow for a possibility is not necessarily to examine it. We noted in Chapter 5 that a heuristic may prune the search-tree to manageable proportions (so avoiding a brute-force examination of all possibilities), while ensuring that the solution is not missed. ID3 is such a heuristic. Whereas its processing is a strictly ordered sequence of decisions, human thinking is not.

This difference is interesting, and important. It reduces the program's psychological relevance. But it does not prevent a program from *appearing* to be creative (the crux of the second Lovelace-question). To deny a system's toehold in creativity because of it would be unreasonable. (Suppose we discovered that the human chess-master used mental processes equivalent to 'logical' heuristics, but unconsciously and in parallel: what then?)

Even so, the inductive programs described so far appear to be creative only to a limited extent. They make new (P-novel) connections, sometimes leading to useful new (H-novel) knowledge. But, although they can restructure the conceptual space (making it easier, for instance, to locate a given soybean disease), the dimensions of that space remain unchanged.

The precise relevance of leaf-spot haloes, or of the white king's being on an edge, may be better appreciated – by us as by them – after they have done their work. But the human specialists already knew, or at least suspected, that haloes and edges were significant. That is why these features were provided to the programs in the first place. No one was led to exclaim: 'Haloes? What could haloes possibly have to do with it?' In short, the ID3 algorithm cannot generate *fundamental* surprises.

In saying this, we must be careful to remember just what is at issue. After all, a person faced with the geometrical problem (in Chapter 5) about isosceles triangles might say 'Congruence? What could congruence possibly have to do with it?' But the geometry-program, which used congruence to solve the problem, was even less creative than the

ID3 algorithm (and much less creative than Pappus of Alexandria, as we have seen). It surprised its programmer, and it surprised us. Indeed, given our human (visual) way of thinking about geometry, these count as fundamental surprises. But the geometry-program did not – so to speak – surprise itself.

Likewise, an ID3-program cannot surprise itself. It cannot produce a fundamental change in its own conceptual space. A more creative program (such as some discussed later) would be able to do this, and might even be able to recognize that it had done so.

The motto of science, today, seems to be 'publish or perish'. The H-creativity of scientists, not to mention their employability, is judged by whether their work is published in refereed journals. On this criterion, computers apparently can be creative. Some new ideas generated by a biochemistry-program have been published in the journal of the American Chemistry Society, and the program itself – called meta-DENDRAL – was given a 'byline' in the paper's title.[4]

To publish a scientific paper, however, one need not be an Einstein. Kuhnian puzzle-solving, of a not very exciting kind, is all that is required. And that is the most that meta-DENDRAL can manage.

DENDRAL (and its 'creative' module, meta-DENDRAL) was one of the first expert systems, initiated in the mid-1960s and continually improved since then. It is modelled on human thinking to some extent, for it embodies inductive principles first identified by philosophers of science. But it employs some very non-human methods too, such as exhaustive search through a huge set of possibilities. Moreover, its creativity is strictly limited to a highly specialized domain.

The program's chemical knowledge concerns a particular group of complex organic molecules, including some steroids used in contraceptive pills. Specifically, it knows how these molecules behave when they are broken into fragments (by an electron-beam) inside a mass spectrometer.

Just as Kekulé already knew the number of carbon and hydrogen atoms in benzene, so modern analytical chemists generally know what atoms make up a given compound. But they may not know just how they are put together. That is, they know the components without knowing the structure. In general, chemical theory allows for many possible structures – often, many thousands. Finding the right one, then, is not a trivial problem.

Because molecules break at 'weak' points, chemists can often analyse an unknown compound by breaking it up and identifying the various fragments. DENDRAL is designed to help them do this. It suggests ideas about a compound's molecular structure on the basis of its spectrograph (the record of fragments), and also indicates how these hypotheses can be tested.

In addition, the program maps all the possible molecules (of a few chemical families) for a given set of atoms, taking account of valency and chemical stability. And it examines this map, using chemical heuristics to identify potentially 'interesting' molecules which chemists might then decide to synthesize.

Originally, DENDRAL had to rely entirely upon its programmers to supply it with chemical rules about how compounds decompose. But a further module (meta-DENDRAL) was added, to find new rules for the base-level program to use. In other words, meta-DENDRAL explores the space of chemical data to find new constraints, which transform (enlarge) the conceptual space that DENDRAL inhabits.

In searching for new rules, meta-DENDRAL identifies unfamiliar patterns in the spectrographs of familiar compounds, and then suggests chemically plausible explanations for them.

For instance, if it discovers that molecules of certain types break at certain points, it looks for a smaller structure located near the broken bonds. If it finds one, it suggests that other molecules containing the same sub-molecular structure may break at these points, too. Similarly, it tries to generalize newly observed regularities in the way that atoms migrate from one site in a molecule to another. Some of its hypotheses turn out to be false, but they are not chemically absurd: none is 'a tissue of [chemical] fancies'.

This program is creative, even H-creative, up to a point. It not only explores its conceptual space (using heuristics and exhaustive search) but enlarges it too, by adding new rules. It generates hunches (about 'interesting' molecules) that human chemists can check. It has led to the synthesis of a number of novel, chemically interesting, compounds. It has discovered some previously unsuspected rules for analysing several families of organic molecules. It even has a publication on its *curriculum non-vitae*.

However, it is limited to a tiny corner of biochemistry. And it relies on highly sophisticated theories built into it by expert chemists (which is why its hypotheses are always plausible). It casts no light on how those theories arose in the first place. Where did today's chemistry come from?

Human chemists remembered for their H-creativity include Johann Glauber, Georg Stahl and John Dalton. Their names are associated with important scientific discoveries (although other people contributed to these discoveries too).

Glauber, in the mid-seventeenth century, clarified the distinction between acids, alkalis, bases and salts. Stahl, in the eighteenth century, helped to show how to discover which elements make up a given compound. He also developed the phlogiston theory of combustion, which was plausible at the time but was later displaced by the theory of oxygen. And Dalton (in 1808) showed that all substances (elements and compounds) consist of individual particles – as opposed to continuous stuff.

Each of these theories was relatively general (no tiny corners of chemistry, here). And they were increasingly fundamental: from qualitative differences between different classes of substance, through the principles of componential analysis, to atomic theory.

Glauber, Stahl and Dalton were all influenced by Francis Bacon's ideas about how to think scientifically. Early in the seventeenth century, Bacon had suggested methods for inducing general laws from empirical data. Other H-creative scientists, too, worked in the Baconian tradition, some of whom are remembered for the fundamental discoveries they made: Joseph Black (who formulated the law of specific heat), Georg Ohm (the discoverer of electrical resistance), Willebrord Snell (who originated the law of refraction), Robert Boyle (who found the first of the gas laws) and many others.

This is hardly surprising, for it was Bacon's writings – together with those of Descartes – which gave rise to modern science as we know it. (You can see the dramatic effect of this revolution in the study of nature by reading Joseph Glanvill's *The Vanity of Dogmatizing*, a pamphlet originally written in the older style but then rewritten in the newly scientific fashion.)[5]

Bacon insisted that science is data-driven, that scientific laws are drawn from experimental observations. We now recognize that science is not *merely* data-driven, since our theories suggest which patterns to look for (and which experiments to do). But Bacon's basic insight stands: scientists do search for regularities in the experimental data. Moreover, if the relevant theoretical framework is not yet established, they can have only the sketchiest notion of what they hope to find. In such cases, they explore the data in a relatively open-ended way.

Data-driven scientific discovery has recently been modelled in a suite of several closely related computer programs.[6] Written with human beings very much in mind, they draw on ideas from the philosophy of science, the history of science (including detailed laboratory-notebooks) and psychology.

The senior member of the design-team, Herbert Simon, is based in a psychology department. As a young man, he was a student of the philosopher of science Carl Hempel. Later, he originated some of the core-concepts of artificial intelligence – such as *search*, *search-space*, *heuristic*, *planning*, *means–end analysis* and *production system* (most of which are crucial to the programs in the suite described below). His pioneering work on human problem-solving has given us new theories, and many ingenious psychological experiments – including some designed specifically for this project. For good measure, he is a Nobel-prizewinner (in economics), too. Clearly, as well as knowing what scholars have written about creativity, he knows what it is like to be H-creative oneself.

The inductive programs inspired by him are intended not to do useful things for working scientists, but to throw light on the nature of scientific creativity as such. As we shall see, they focus on the consciously accessible aspects of scientific thinking, rather than the tacit recognition of patterns and analogies.

They have rediscovered (P-created) many important principles of physics and chemistry, including Black's law for example. And they are called – yes, you guessed it! – BACON, BLACK, GLAUBER, STAHL and DALTON.

What BACON does is to induce quantitative laws from empirical data. It is given sets of numbers, or measurements, each set recording the values of a certain property at different times. Using a variety of numerical heuristics, it searches for mathematical functions relating the property-values in a systematic way. In other words, what it finds scientifically 'interesting' are invariant relations between (numerical) data-sets.

The first question BACON asks is whether the corresponding measurements are directly or inversely proportional, and if so whether there are any constants involved in the equation relating them. If the program finds no such function relating the two sets of measurements directly, it introduces a new theoretical concept defined in terms of them. Then, it can look for a principle involving the newly coined term.

For instance, BACON can multiply the corresponding values of the two properties by each other, so defining their *product*, and then consider that. Perhaps the product is a constant, or is systematically related to a third property? (This third property may also be a theoretical construct, defined in terms of observables.) Again, the program can divide one value by the other to explore the *ratio* of the two data-sets. Or it can multiply each value by itself, to look for a *power-law*. If necessary, it can use several of these numerical heuristics in relating the two measurement-sets.

Using only these relatively simple rules, the program rediscovered a number of important scientific laws. For instance, it came up with Boyle's law relating the pressure of a gas to its volume ($PV = c$), and a version of Ohm's law of electrical resistance ($I = v/(r + L)$. Ohm's law is more complex than Boyle's law, because there are two constants involved (namely, v and r). BACON noticed that the current passing through a wire decreases as the wire's length increases, so asked itself whether their product (LI) is constant. It isn't. But it is related in a fairly simple (linear) way to the values of the current itself, and BACON realized that fact. (A later version of the program expressed Ohm's law by means of the more familiar equation, as we shall see.) These primitive inductive methods also enabled BACON to derive Galileo's law of uniform acceleration, which states that the ratio of distance to the square of the time is constant ($D/T^2 = k$). And it P-created Kepler's third law, which defines a constant ratio between the *cube* of the radius of a planet's orbit and the *square* of its period of revolution around the sun (D^3/P^2).

(The program P-created Kepler's law twice. The first time, it had to use data 'doctored' to make the sums come out exactly right, because it would have been irredeemably confused by the messiness of real data. But an improved version was later able to cope with real data: the very same figures used by Isaac Newton, when he checked Kepler's third law. As human scientists know only too well, all actual measurements are imprecise. So BACON does not now demand mathematical purity. When looking for 'equal' values of a certain property, it can ignore small differences – *how* small, according to the programmer's choice. Consequently, it can tolerate noise in experimental results.)

BACON tries to make its life as easy – and its science as elegant – as possible, by looking for the most obvious patterns first. In selecting which heuristic to use, it does not doggedly run through a list, one by one. Rather, it considers them all 'in parallel', giving priority to the simplest one that is applicable in the particular case. Nevertheless, it is a sequential system, since no heuristic can be brought into play until the previously chosen heuristic has completed its work. The analogy is with a human scientist who tries one thing after another, always trying the simplest possibilities first.

The heuristics at the heart of the earliest version of the program, in order of priority, were:

IF the values of a term are constant, THEN infer that the term always has that value.
IF the values of two numerical terms give a straight line when plotted on a graph, THEN infer that they are always related in a linear way (with the same slope and intercept as on the graph).
IF the values of two numerical terms increase together, THEN consider their ratio.
IF the values of one term increase as those of another decrease, THEN consider their product.

With these heuristics, BACON can discover only laws that are very close to the data (laws which can be restated in terms of observables). But BACON exists in five different versions, equipped with heuristics of increasing mathematical power. These can construct theoretical concepts whose relation to the experimental results is much less direct.

The later versions of BACON can explore, construct and transform conceptual spaces of increasing complexity and depth. They can define second-level theoretical terms, by using (for instance) the *slope* and *intercept* referred to in the second heuristic, above. Indeed, they can construct concepts at successively higher levels, each defined in terms of the theoretical concepts of the level below. And they can find relations between laws, not just between data or theoretical constructs.

Moreover, they can relate more than two sets of measurements. They can cope with inaccurate data, up to a point. They can cope with irrelevant data, at first investigating all the measurable properties, but later ignoring those which turn out to be of no interest because they show no regular variation. They can suggest experiments, to provide new sets of correlations, new observations, with which they then work.

They can also introduce new basic units of measurement, by taking one object as the standard. (Human scientists often choose water.) And they can use the notion of symmetry, as applied to equations, to help them to find invariant patterns in the data.

The maturer BACON has come up with many scientifically significant P-creations. The third version, for example, discovered the ideal gas laws ($PV/t = k$). It even 'reinvented' the Kelvin temperature-scale, by adding a constant of 273 to the Celsius value in the equation.

The fourth version constructed Ohm's notions of voltage and resistance, and expressed his discovery in the familiar form ($I = v/r$). It emulated Archimedes, in discovering the law of displacement relating volume and density. (Admittedly, it was given the hint that objects could be immersed in known volumes of liquid, and the resulting volume measured.) It also emulated Black, finding that different substances have different specific heats. (A substance's specific heat is the amount of heat required to raise the temperature of one gram of it from 0°C to 1°C.)

In addition, the fourth version discovered – at a purely descriptive level – the concepts of atomic and molecular weight. In other words, it looked for *small integer ratios* between the combining weights and volumes of chemical substances, and often picked out what we know to be the correct atomic and molecular weights. However, it sought no explanation for these numbers (in terms of tiny individual particles, for instance).

The fifth incarnation of BACON included a symmetry-heuristic, applicable to equations, which it used to find Snell's law of refraction. Also, it produced a version of Black's law which is more general than the one produced by BACON-4. (More accurately, BACON defines the *reciprocal* of specific heat. Its equation was therefore inelegant, although mathematically equivalent to Black's law. The programmers suggested an extra heuristic, which would enable the program to come up with the more familiar equation, and which could simplify many other equations too.)

Black, however, did not derive his law of specific heat from the experimental measurements alone. As well as being data-driven, he was theory-driven too. That is, he was partially guided by a hunch that the quantity of heat would be conserved.

The quantity of heat is not the same thing as the temperature. Indeed, Black's theory clarified this difference, which is a special case of the distinction between *extensive* and *intensive* properties. Extensive properties are additive, but intensive properties are not. Mass is extensive: if you add 1 gram of water to 100 grams, you get 101 grams. Temperature is intensive: if you add boiling water to icewater you do not get water at 101 degrees Centigrade. All conservation laws concern extensive properties, since they state that the total quantity of something remains constant throughout the experiment.

Science has many conservation laws, and a scientific-discovery program should be able to find them. Simon's team wrote another program accordingly. It can be thought of as an extension of BACON (and might be added as a special module), because – like BACON – it finds quantitative laws unifying numerical data. But the programmers gave it a different name (BLACK), to mark the fact that it is more theory-driven than BACON.

BLACK considers situations in which two objects combine to form a third (for example, hot and cold water mixed together). In doing so, and in defining new theoretical terms, it uses heuristics that distinguish between extensive and intensive properties. If the measurements show that all the observable properties are additive, and therefore extensive, BLACK has no work to do. But if the data show that some property – temperature, for example – is not extensive, the program tries to find conservation laws accounting for these non-additive data in terms of some newly defined extensive property.

In this way, BLACK came up with a (third) statement of the law of specific heat. BACON had P-created a theoretical term (the reciprocal of specific heat) to *summarize* the experimental results. But BLACK's version of Black's law *explains* the data, by hypothesizing that an unobservable property (quantity of heat) is conserved.

You may be muttering: 'There is more to nature than numbers!' You are right. Science is not all about equations. Moreover, even when we do have an equation we want to know why it is true, we want to be able to explain it – in terms of a structural model, perhaps.

So Simon and his colleagues have designed three more programs, all dealing with non-quantitative matters. Like BACON, they are strongly data-driven. But, like BLACK, they have very general 'hunches' about what they can expect to find. Specifically, they tackle the sorts of problems originally mapped by Glauber, Stahl and Dalton.

GLAUBER discovers qualitative laws, laws that summarize the data by classifying things according to their observable properties. Such properties include a substance's taste and colour, and its behaviour in a test-tube.

Qualitative laws are needed because scientists cannot always measure a property they are interested in. Indeed, qualitative laws are often discovered long before they can be expressed numerically. For instance, people already knew that babies, animals and plants inherit certain traits from their ancestors when Mendel discovered quantitative laws of inheritance (stating the average ratios of different traits in the offspring).

The chemist Glauber clarified the qualitative distinction between acids and alkalis, and between acids and bases (bases include both alkalis and metals). He did this by classifying the experimental observations – including some produced in experiments he designed himself – in a logically coherent way. He discovered, for instance, that every acid reacts with every alkali to form some salt.

To do the same sort of thing, GLAUBER uses the branch of logic which deals with statements saying, 'There exists a particular substance with such-and-such properties,' or 'All the substances in a certain class have such-and-such properties.' The program applies this logic to facts about the observable properties and reactions of chemical substances.

It can be told, for example, that hydrochloric acid tastes sour, and that it reacts with soda to form common salt. Given a number of such facts, GLAUBER can discover that there are (at least) three classes of substance: acids, alkalis and salts, and that these react with each other in a regular fashion. Moreover, it can define higher-level classes (such as bases), and then make hypotheses about those classes too.

Like Glauber, GLAUBER does not insist on investigating all the acids in the universe – or even all the acids it knows about – before forming the generalization that 'all' acids react with alkalis to form salts. It does, however, test its hypotheses by ensuring that *most* of the acids it knows about have been observed to react in this way. Like human beings, then, it can tolerate missing evidence. If this sort of fuzziness were not possible, scientific reasoning (induction) could never get off the ground.

Unlike human beings, however, GLAUBER cannot deal with counter-examples, or *negative* evidence. Nor can it design new experiments, in an attempt to disconfirm its hypotheses. The reason is that the program's logic cannot distinguish between *denying* a statement and *not asserting* it. This logical difference is widely respected by people. It is crucial not only to experimental method, but to everyday tactfulness as well: to tell a friend that her new hat does not suit her is not the same as avoiding the topic altogether.

An improved version of GLAUBER is being developed, to overcome this limitation, and others. Meanwhile, the program's grasp of experimental method is much less powerful than ours.

Human scientists try to provide explanations, as well as descriptions. BACON and GLAUBER can offer only descriptive summaries of the data. BLACK dips its toes into the waters of explanation, by postulating the conservation of underlying properties. But STAHL and DALTON go further: the one wading in up to its ankles, while the other gets its shins thoroughly wet.

STAHL analyses chemicals into their components, saying what elements make up a certain substance. Like Stahl himself, it does not say whether elements are made of separate particles of continuous stuff, nor identify the proportions of the elements involved. (DALTON takes a stand on these issues, as we shall see.)

The input to STAHL is a list of chemical reactions observed in the laboratory. Each reaction is described by saying that when *these* substances reacted with each other, *those* substances were observed to result. The program's output is a list of substances described in terms of their components. It learns as it goes, for it remembers its previously derived componential analyses and uses them in reasoning about later inputs.

This program does not model an isolated flash of scientific insight, or even an afternoon's creative work. Rather, it models the publicly argued progress of science over many years. If STAHL is given experimental results in the order that they were observed in history, it comes up with the explanatory theories – sometimes mistaken, but always plausible – put forward by chemists in centuries past: not only Stahl, but also Henry Cavendish, Humphry Davy, Joseph Gay-Lussac and Antoine Lavoisier.

For this to happen, the experimental data must be input to STAHL in the way they were described at the time. For example, the program may be told that when charcoal is burnt in air, the result is ash, phlogiston and air. This is how the reaction was reported early in the eighteenth century.

The phlogiston theory stated that combustible substances contain phlogiston, which is given off when they burn. Phlogiston was believed to be visible as the fire observed in experiments on combustion. Stahl originated the phlogiston theory in about 1700. The theory was further developed for almost a century, being successively adapted to fit the new experimental data as they emerged. (For example, when it became clear that many substances become heavier on burning, phlogiston was declared to have *negative* weight.) It was not until the 1780s that Lavoisier's oxygen theory was widely accepted. Many steps of this progression through a changing theoretical space have been replicated by STAHL, using the experimental data input in their historical order.

Simon's team believe that human scientists use similar methods of reasoning in arguing for different, or even competing, theories. Stahl and Lavoisier, for example, thought in essentially similar ways; their differences lay in the experimental data available to them, and the theoretical assumptions from which they started. Accordingly, STAHL's programmers ensure that it always uses the same reasoning methods – the same set of heuristics.

The heuristics used by STAHL represent forms of reasoning which chemists (as their notebooks show) have used time and time again. The three basic rules used to derive STAHL's componential analyses are:

INFER-COMPONENTS:
IF A and B react to form C,
OR IF C decomposes into A and B,
THEN infer that C is composed of A and B.

REDUCE:
IF A occurs on both sides of a reaction,
THEN remove A from the reaction.

SUBSTITUTE:
IF A occurs in a reaction,
AND A is composed of B and C,
THEN replace A with B and C.

The program also has two heuristics that enable it to discover that two differently named substances are one and the same. One of these rules applies when STAHL notices that a substance (A) decomposes in two different ways, with the decompositions differing by only one substance:

IDENTIFY-COMPONENTS:
IF A is composed of B and C,
AND A is composed of B and D,
AND neither C contains D nor D contains C,
THEN identify C with D.

The fifth heuristic enables STAHL to conclude that two different substances are one and the same because they are compounds made up of the same components:

IDENTIFY-COMPOUNDS:
IF A is composed of C and D,
AND B is composed of C and D,
AND neither A contains B nor B contains A,
THEN identify A with B.

(If you think about that one for a moment, you will spot a trap in it: we now know that two distinct compounds *may* be composed of just the same elements, present in different proportions or structured in different ways.)

Reactions are described to STAHL by using the old-fashioned chemical names: *air*, *charcoal*, *sulphur*, *iron*, *calx of iron*, *ash*, *soda*, *potash*, *vitriolic acid*, *muriatic acid*, *litharge*, *lime*, *quicklime*, *phlogiston*. STAHL must use the observational data to derive hypotheses about the chemical make-up of the various substances involved. This problem is not a straightforward one. Indeed, it is likely to lead to blind alleys and false trails – some of which may be followed, and elaborated, for some time before their inadequacy is realized. For the terms in which Stahl and his near-contemporaries reported their experiments are not merely archaic. They betray some false assumptions, and fail to mark some important distinctions.

For example, 'phlogiston' – the word used in reporting the observation of fire – does not refer to an actual substance at all. Again, we now know that 'air' is not (as early theorists assumed) a single substance. Rather, it is a mixture of nitrogen, oxygen and carbon-dioxide – with traces of other gases besides. If experiments are reported in language riddled with false assumptions, mistakes are likely to result when scientists try to explain their observations in componential terms.

One can expect trouble, for instance, when the REDUCE heuristic is used to delete references to 'air' on both sides of a reaction. This line of reasoning misled eighteenth-century chemists into ignoring air – and therefore any components it might have – when explaining the reaction (described above) involved in the combustion of charcoal. Similar reasoning, involving not air but water, led chemists a century later to see sodium as a compound of soda and hydrogen. In other cases, the REDUCE heuristic led people to mistake an element for a compound: iron, in Stahl's time, was thought to be a compound of calx of iron (iron ore) and phlogiston.

It does not follow that the REDUCE heuristic should be dropped, for it is a very useful method of thinking. So are the other heuristics, even though they sometimes lead STAHL astray. (In this, they are no different from heuristics in general.) But it does follow that STAHL needs some way of recovering from its mistakes. Some mistakes arise not because a heuristic is faulty, but because the heuristics are not always applied in the same order. Consequently, STAHL can arrive at different conclusions from the very same evidence. The errors it makes as a result are often errors made in the past by human chemists, some of which remained uncorrected for many years.

Human chemists typically identify their errors by finding an inconsistency in their thinking. Often, they can correct it by using chemical knowledge that was not available when the mistake was first made. STAHL, too, can use its recently acquired knowledge to identify and correct faulty reasoning.

The program can realize, for instance, that it has assigned two different componential models to one and the same substance. And it can recognize an especially over-enthusiastic use of the REDUCE heuristic: one which results in a reaction with no input, or no output. It deals with such inconsistencies by reconsidering all the reactions involving the substances concerned, taking into account all of the componential analyses it has generated so far (except the inconsistent pair). Very likely, some of these analyses were not yet available when the program first came up with the model now causing the trouble. Since the program knows more now than it did before, it is not surprising that this approach often gets rid of the inconsistency.

Sometimes, STAHL finds itself running around a logical circle. For example, it may have decided that substance A is composed of B and D, *and* that substance B is composed of A and D. When it realizes that this has happened, it introduces a new distinction (a new name for one of the As), and rewrites one of the troublesome statements accordingly. (It chooses the statement which is further from the input data, the one derived by the more complex reasoning.) This resembles historical cases where human chemists have suggested, with no specific evidence besides a circularity, that a substance in a certain reaction must have been impure.

The dimensions of STAHL's conceptual space, and the ways in which it wanders through it, are intriguingly close to the human equivalents. But there is much which STAHL (in its current version) cannot do.

It cannot deal with quantities, as BACON can (so it cannot ask whether phlogiston might have a negative weight). Unlike GLAUBER, it cannot reason about qualities (so it cannot consider Lavoisier's hypothesis that substances containing oxygen have an acid taste). It cannot design new experiments, as some versions of BACON can (so it suggests no new tests of its analysis of iron into phlogiston and calx of iron). Because it always chooses the 'best' componential model and throws away the others it had been thinking about, it cannot compare two competing theories (phlogiston and oxygen, for example).

Moreover, it knows nothing of atoms, or molecular structure. Those are DALTON's concerns.

Chemists (Kekulé, for instance) may know the components of a substance without knowing its structure. They have to discover how many atoms, of which elements, combine to form a single molecule, and how those atoms are linked to each other. Clearly, these questions cannot even be asked without assuming atomic theory. DALTON takes (an early version of) atomic theory for granted, and uses it to generate plausible molecular structures for a given set of components.

Its input is the type of information that STAHL produces as output: lists of chemical reactions, with each substance described in terms of its components. It can be told, for instance, that water is composed of hydrogen and oxygen, both of which are elements, and that hydrogen reacts with nitrogen (also an element) to produce ammonia. Its output lists the atoms, and their proportions, in the molecules of the substances concerned. It cannot say which atom is linked to which (Dalton did not specify neighbour-relations, either).

DALTON is of no use to working chemists, as DENDRAL is, for it generates no H-creative ideas. Rather, it suggests how early atomic theory provided scientists with primitive ways of mapping, and exploring, the space of molecular structures.

For instance, atomic theory implies that the *number* of atoms in a molecule, or of molecules in a reaction, is important – and the program's heuristics take this into account. If DALTON is not told the numbers of molecules of the various substances in a reaction (being informed merely that 'hydrogen reacts with oxygen to form water'), the SPECIFY-MOLECULES heuristic assumes that each of these numbers is some small integer (from 1 to 4). This allows for a finite set of candidate-equations for the reaction in question.

Similarly, the SPECIFY-ELEMENTS heuristic tells the program to assume that the number of atoms in one molecule of any element can range from 1 to 4. (Dalton himself did not use this heuristic, for he believed that atoms of the same element repel each other, so that molecules of elements must be monatomic; as a result, he insisted that water must be analysed as HO, not as H_2O.)

Atomic theory implies also that, in any chemical reaction, the total number of atoms of each element remains the same. The SPECIFY-COMPOUND heuristic exploits this fact. It considers the set of candidate-equations allowed by the two rules described above. Each equation specifies some number of molecules for every substance in the reaction. Provided that only one of the substances lacks a structural description, this heuristic can assign one to it by using the conservation principle just defined. (Successful analyses are stored, so a problem that is insoluble today may be solved tomorrow.)

DALTON prefers simple analyses to complex ones. So SPECIFY-COMPOUND always considers the low-numbered equations first: the 1s before the 2s, and the 4s last of all. (Dalton himself recommended the same strategy.)

DALTON's designers suggest that it might one day be extended, to deal (for example) with elementary-particle physics or Mendelian genetics. These extensions, however, would involve some fundamental changes.

Meanwhile, they plan to integrate BACON, BLACK, GLAUBER, STAHL and DALTON into one system. The five programs are constructed in the same sort of way, and tackle nicely complementary problems, so the output of one might function as the input to another. The result would be a model able to explore conceptual spaces of many different kinds, in many different ways.

Just how creative do these five programs appear to be, and how closely do they model human creativity?

They range over a wide field of science, as opposed to burrowing in little holes of specialist expertise. They have P-created many important scientific laws, and many plausible hypotheses. And they do so in ways which, to a significant degree, match the ways in which individual scientists generate and justify their ideas. (Simon's group provide a wealth of detailed historical evidence for this claim.)

However, they have made no H-novel discoveries (as DENDRAL has). To be sure, BACON has come up with new formulations of well-known laws – involving the reciprocal of specific heat, for instance. But its novel formulations are no more elegant than the familiar versions, and are sometimes inferior.

We can ignore (for example) STAHL's inability to deal with measurements. For BACON is not limited in this way, and an integrated discovery-program could pool the reasoning-methods of all five programs. Even so, there are many things these programs cannot do.

For instance, their names are mere courtesy-titles: they had to be provided with the ways of reasoning which Bacon, Glauber, Stahl and Dalton had pioneered before them. Admittedly, all human scientists take advantage of methods pioneered by others. A few, however, can alter – and make explicit – the general principles of reasoning with which to approach the task. The BACON family cannot do that. They can learn: they can add many details to the maps of their conceptual space, and so follow pathways they did not know about before. But they cannot make fundamental changes in the nature of the space.

Moreover, the five discovery-programs have to be given the concepts they are to think about. For instance, BACON-4 was able to generate Archimedes' principle only because it had been told that things could be immersed in known volumes of liquid, and the resulting volume measured. To be sure, novel insights like the one that caused Archimedes to leap out of his bath are relatively rare. Much H-creative science is either puzzle-solving, where the crucial concepts are already known, or exploration guided by hunches (which may be widely shared at the time) that familiar concepts are somehow relevant. Nevertheless, it is fair comment that BACON *et al.* are, at base, utterly dependent on concepts provided by the programmers. (An integrated version would be less open to this criticism, since one program would feed its output into the next.)

These programs are limited, too, in that they model only the consciously accessible aspects of scientific discovery. They are concerned with the sort of thinking which, even if it is not carried out consciously at first, can at least be deliberately checked and justified. In short, they offer us a theory of scientific *reasoning*. They do not model the sudden flashes of insight often reported by H-creative scientists.

Many of these insights involve the recognition of patterns or analogies (between snakes and curves, for instance), as opposed to careful reasoning like that of the BACON tribe. Such insights might be modelled by a different type of computational system, more like ARCS and ACME. Future computer-scientists may engage in both types of thinking: sequential–deliberative and parallel–intuitive. But as we saw in Chapter 6, it is not yet clear how the two sorts of processing can be combined.

Finally, these programs know not of what they speak. They would not know a test-tube if they saw one – indeed, they cannot see one. They have no cameras to pick up rising water-levels, and no bodies to lower into the bath. They do not react to heat and cold, so their concept of temperature is 'empty'. Some of them can plan experiments, but none can do them. In short, they have virtually no causal connections with the material world, responding only to the fingers of their programmers tapping on the teletype.

However, BACON and friends are not intended to be robot-scientists. Rather, they focus on certain abstract conceptual structures – mathematical functions, classifications, componential analyses – which map the spaces of human thought.

When discussing the necklace-game in Chapter 4, I said that it gave a flavour of what it is like to do creative mathematics. This involves not doing sums, but combining and transforming ideas, and exploring analogies between them. The necklace-game showed how one can play around with mathematical rules and concepts to see if anything interesting crops up – subtraction, for example, or *every* lucky number.

Creative mathematics is the focus of Douglas Lenat's 'Automatic Mathematician', or 'AM'.[7] This program does not produce proofs (it does not model the 'verification' phase). Rather, it generates and explores mathematical ideas, coming up with new concepts and hypotheses to think about.

AM starts out with a hundred very primitive mathematical concepts. (They are drawn from set-theory, and include sets, lists, equality and operations). These concepts are so basic that they do not even include the ideas of elementary arithmetic. To begin with, the program does not know what an integer is. As for addition, subtraction, multiplication and division, these are as unknown to the infant AM as the differential calculus is to the child playing the necklace-game.

Also, AM is provided with about 300 heuristics. These can examine, combine and transform AM's concepts – including any compound concepts built up by the program. Some are very general, others specific to set-theory (these include some specializations of the more general ones). They include ways of comparing concepts in any domain, togther with some tricks of the mathematician's trade, and they enable AM to explore the space potentially defined by the primitive concepts. This involves both conceptual change (combinations and transformations) and enquiries aimed at 'mapping' the domain.

Among the map-making questions that AM can ask about a concept are these: Is it named? (The human user can nudge AM in certain directions, by giving a newly defined concept a name: AM is more likely to explore a concept if it is named.) Is it a generalization or a special case of some other concept? What examples fit the definition of the concept? Which operations can operate on it, and which can result in it? Are there any similar concepts? And what are some potential theorems involving the concept? (Some of these questions, as you may have noticed, are similar to those used by ACME and ARCS for exploring analogies in natural language.)

Among the transformations that AM can carry out on a concept is to consider the *inverse* of a mathematical function. This heuristic is a mathematical version of our old friend, *consider the negative*. It enables the program (for example) to define multiplication having already defined division, or to define square-roots having already defined squares. Another transformation generalizes a concept by changing an 'and' into an 'or' (compare relaxing the membership-rules of a club, from 'Anyone who plays bridge and canasta' to 'Anyone who plays bridge or canasta').

However, AM does not blithely consider *every* negative, nor change *every* 'and' into an 'or'. Time and money do not allow it to ask every possible question, or change each concept in all possible ways. So which concepts does it focus on, and which changes does it actually try?

Like all creative thinkers, AM needs hunches to guide it. And it must evaluate its hunches, if it is to appreciate its own creativity. A program that rejected every new idea as 'a cartload of dung' would not get very far. Accordingly, some of AM's heuristics suggest which sorts of concepts are likely to be the most interesting. If it decides that a concept is interesting, AM concentrates on exploring that concept rather than others. (If it regards a concept as *uninteresting*, it can try to make it more noteworthy by changing it in various ways – by generalization, for instance.)

What is interesting in mathematics is not – or not entirely – the same as what is interesting in dress-design, or jazz. If an operation can be *repeated* an arbitrary number of times, that fact (as AM knows) is mathematically interesting. It is musically interesting too, with respect to a twelve-bar blues. It is chemically interesting, since many organic molecules are based on long strings of carbon atoms. It is grammatically interesting, to an author writing about 'a squeezing, wrenching, grasping, scraping, clutching, covetous old sinner'. It is relevant to couturiers, who can use it to design multiply flounced skirts (like those worn by flamenco dancers). It is even useful to a ten-year-old trying to draw 'a funny man'.

Some other features that mathematicians find interesting are more arcane. This is only to be expected, since creative thinking involves expertise. A dress-designer who knows nothing about flounces is professionally incompetent, but a similarly ignorant musician is not. But even some 'arcane' features are less domain-specific than they may seem. For instance, AM takes note if it finds that the union of two sets has a simply expressible property that is not possessed by either of them. This is a mathematical version of the familiar notion that *emergent* properties are interesting. In general, we are interested if the combination of two things has a property which neither constituent has.

Like human hunches, AM's judgments of what ideas are most promising are sometimes wrong. Nevertheless, it has come up with some extremely powerful notions.

It produced many arithmetical concepts, including *integer*, *prime*, *square root*, *addition* and *multiplication* – which it notices can be performed in four different ways, itself a mathematically interesting fact. It generated, though did not prove, the fundamental theorem of number theory: that every number can be uniquely factorized into primes. And it came up with the intriguing idea (known as Goldbach's conjecture) that every even number greater than two is the sum of two different primes.

On several occasions, it defined an existing concept of number theory by following unusual paths – in two cases, inspiring human mathematicians to produce much shorter proofs than were previously known. It has even originated one minor theorem which no one had ever thought of before. This theorem concerns a class of numbers ('maximally divisible' numbers, first described in the 1920s) which Lenat himself knew nothing about.

In short, AM appears to be significantly P-creative, and slightly H-creative too. As with the Pappus-program, however, AM's creativity can be properly assessed only by close examination of the way in which it works.

Some critics have suggested that its performance is deceptive, that many of the heuristics may have been specifically included to make certain mathematical discoveries possible.

In reply, Lenat insists that the heuristics are fairly general ones, not special-purpose tricks. On average, he reports, each heuristic was used in making two dozen different discoveries, and each discovery involved two dozen heuristics. This does not rule out the possibility that a few heuristics may have been used only once, in making an especially significant discovery. (A detailed trace of the actual running of the program would be needed to find this out.) If so, the question would then arise whether they had been put in for that specific purpose, as opposed to being included as general methods of exploring mathematical space.

Lenat does admit, however, that AM uses a very helpful representation: it is written in a programming-language (LISP) whose syntax is well suited to model mathematics. Heuristics that make small 'mutations' to the syntax of LISP-expressions are therefore quite likely to come up with something mathematically interesting. In short, writing AM in LISP is equivalent to building in, implicitly, some powerful mathematical ideas. If these ideas are different from the ones created by AM, all well and good. If not, 'created' hardly seems the right word.

Moreover, AM was often encouraged by Lenat (sometimes on the advice of professional mathematicians) to focus on certain new ideas rather than others. Had it been left purely to its own devices, the proportion of actually trivial ideas defined and explored by it would have been higher.

The precise extent of AM's creativity, then, is unclear. But we do have some specific ideas about what sorts of questions are relevant. Without doubt, it is a stronger candidate for mathematical creativity than the geometry-program discussed in Chapter 5. (AM can explore geometry, too: Lenat reports that it was 'almost as productive there as in its original domain'.)

The programs discussed so far, in this chapter and the previous one, cannot modify their own processing in a fundamental way. They can explore conceptual spaces, and build many P-novel – and even some H-novel – structures in the process. They can exploit heuristics of many sorts: some focused on melodic contours or theoretical chemistry, others on verbal or mathematical analogy-mappings of highly abstract kinds. But all are 'trapped' within a certain thinking-style.

The reason is that (like the very youngest children described in Chapter 4) they cannot reflect on their own activity. They have no higher-level procedures for changing their own lower-level rules, no 'maps' of their own heuristics with which to guide the exploration or change of those heuristics themselves.

You may object that such guidance is unnecessary for creativity, since many H-novel structures have been produced without it. The creative strategy of evolution, you may say, is *Random-Generate-and-Test*: new biological structures are generated by random mutations, and then tested by natural selection. So randomness, not carefully mapped guidance, is seemingly all that is needed for generating fundamental change. Indeed, you may recall a number of occasions on which human creativity involved mere chance: Fleming's discovery of penicillin, for instance.

It is true that random events sometimes aid creativity (many examples will be given in Chapter 9, where we shall also discuss just what 'random' means). However, creativity in human minds cannot be due only to random changes in pre-existing structures. Biological evolution has had many millions of years in which to generate a wide variety of novelties, and in which to weed out the useless ones. But we must improve our thinking within a single lifetime (or, collectively, within the history of a certain culture or the cumulative experience of the human race).

What we need, then, is not (or not only) Random-Generate-and-Test, but *Plausible-Generate-and-Test*. That is, we need some way of guiding our creative exploration into the most promising pathways. Sometimes we shall be misled, of course. But without some sense of what sort of changes are likely to be fruitful, we would be lost in an endless thrashing about. (Even biological evolution, as we shall see in the next section, relies to some extent on Plausible-Generate-and-Test.) In short, the reasons which make heuristics necessary in the first place also underlie the need for *heuristics for changing heuristics*.

What sorts of processes might these be? Many, no doubt, are highly specific to a given domain, and are part of the expertise which an experienced person possesses. But others will be much more general. Some of these have been modelled in a program called EURISKO (written by Lenat, who also wrote AM), which explores and transforms its own processing-style.[8]

For example, one heuristic asks whether a rule has ever led to any interesting result. If it has not (given that it has been used several times), it is marked as less valuable – which means that it is less likely to be used in future.

What if the rule has occasionally been helpful, though usually worthless? Another heuristic, on noticing this, suggests that the rule be specialized. The new heuristic will have a narrower range of application than the old one, so will be tried less often (thus saving effort). But it will be more likely to be useful in those cases where it is tried.

Moreover, the 'specializing-heuristic' can be applied *to itself*. Because it is sometimes useful and sometimes not, EURISKO can consider specializing it in some way. But what way?

Lenat mentions several sorts of specialization, and provides heuristics for all of them. Each is plausible, for each is often (though not always) helpful. And each is useful in many domains – in dressmaking, for example.

One form of specialization requires that the rule being considered has been useful at least three times. (If a certain method of sewing buttonholes almost never works, it may be sensible to stop using it.) Another sort of specialization demands that the rule has been *very* useful, at least once. (Leather, as opposed to the more common fabrics, may require a special way of making buttonholes.) Yet another insists that the newly specialized rule must be capable of producing all the past successes of the unspecialized rule. (A sewing-machine with a new method of hemming should be able to hem all the dresses that the old one could.) And a fourth version specializes the rule by taking it to an extreme. (Some Spanish dressmaker may have designed the first multiply flounced skirt by applying this specialization-heuristic to her repetition-heuristic; and Mary Quant took skirt-length to an extreme in creating the mini-skirt.)

Other heuristics work not by specializing rules, but by generalizing them. Generalization, too, can take many forms. A heuristic may warn the system that certain sorts of rule-generalization (such as the potentially explosive replacement of 'and' by 'or') should normally be avoided. Still other heuristics can create new rules by analogy with old ones. Again, various types of analogy can be considered.

These ways of transforming heuristics (specialization, generalization and analogy) are comparable to AM's ways of transforming concepts. This is no accident, for Lenat's claim is that a many-levelled conceptual space can be explored by much the same processes on any level.

Certainly, special domains will require heuristics, too. Someone who knows nothing about harmony will not be able to create new harmonic forms, and someone who knows nothing about chemistry cannot suggest new substances for chemists to synthesize.

But a composer who has the unconventional idea of modulating from the home key into a relatively 'distant' key is generalizing the notion of modulation. A chemist who suggests synthesizing a new substance is arguing by analogy with the behaviour of familiar compounds of similar structure. And an artist who decides to draw all limbs as straight-sided figures is specializing the previous rules for limb-drawing. The person's expert knowledge suggests which specific sorts of generalization (or analogy, or specialization) are most plausible in the particular case.

With the help of various packets of specialist knowledge, EURISKO has been applied in several different areas. It has come up with some H-novel ideas, concerning genetic engineering and computer-chip (VLSI) design. Some of its ideas have even been patented (the US patent-law insists that the new idea must not be 'obvious to a person skilled in the art').

For instance, the program designed a three-dimensional computer-chip which enabled one and the same unit to carry out two different logical functions simultaneously. (The unit could act both as a 'Not–And' circuit and as an 'Or' circuit.) EURISKO did this by taking the typical three-dimensional junction shown in Figure 8.2(a), and adding three more parts to it (see Figure 8.2(b)). The general heuristic it used was: 'If you have a valuable structure, try to make it more symmetric.' Human designers favour symmetry too. But they had not thought of doing this, nor of the possibility that a single unit could perform two different functions.

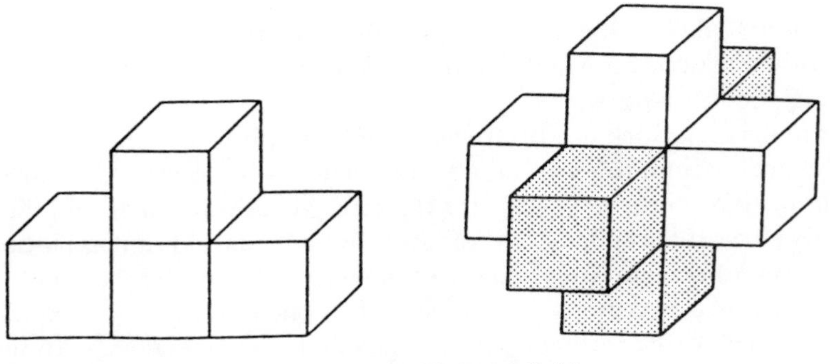

Figure 8.2 (a) **Figure 8.2 (b)**

Moreover, EURISKO has won an official 'creativity-competition' in which all the other contestants were human – indeed, it won it twice. The contest was a war-game, in which one has to design a battle-fleet within certain cost-limits, and then test it (in a simulation) against the fleets of the other players. When EURISKO first played the game, it designed shipping-fleets so unconventional that the human players were convulsed with mirth. Their laughter died when the program won the game. For the next year's competition, the rules were changed to make it harder for EURISKO. Even so, the program won a second time. Then, the rules were changed again: 'No computers'.

A program written a few years ago uses IF–THEN rules to regulate the transmission of gas through a pipeline in an economical way.[9] It receives hourly measurements of the gas-inflow, gas-outflow, inlet-pressure, outlet-pressure, rate of pressure-change, season, time of day, time of year, and temperature. Using these data, it alters the inlet-pressure to allow for variations in demand. In addition, it infers the existence of accidental leaks in the pipeline – and adjusts the inflow accordingly. Moreover, it was not told which rules to use for adjusting inflow, or for detecting accidental leaks. It discovered those rules for itself.

What's so interesting about that? Didn't the soybean-program do the same sort of thing, in learning how to diagnose plant-diseases efficiently? Why discuss yet another example of inductive classification?

The difference is that the pipeline-program discovered its expert-level rules by starting from a set of randomly generated rules, which it repeatedly transformed in part-random, part-systematic ways. It employed a particular form of Plausible-Generate-and-Test, using heuristics called genetic algorithms. These enable a system to make changes that are both plausible and unexpected, for they produce novel recombinations of the most useful parts of existing rules.

As the name suggests, these heuristics are inspired by biological ideas. Some genetic changes are isolated mutations in single genes. But others involve entire chromosomes. For example, two chromosomes may swap their left-hand sides, or their midsections (the precise point at which they break is largely due to chance). If a chromosome contained only six genes (instead of many hundreds), then the strings *ABCDEF* and *PQRSTU* might give *ABRSTU* and *PQCDEF*, or *ABRSEF* and *PQCDTU*. Such transformations can happen repeatedly, in successive generations. The strings that eventually result are unexpected combinations of genes drawn from many different sources.

Genetic algorithms in computer programs produce novel structures by similar sorts of transformation. They are being used to model many sorts of adaptive learning and discovery, of which the pipeline-program is just one example.[10]

Psychological applications of such simple combinatorial methods may seem doomed to failure. Indeed, these very methods are used by Lerner to ridicule the idea of a computer-poet. Almost all the lines in *Arthur's Anthology of English Poetry* (cited in Chapter 1) are derived, by 'mechanical' recombinations, from the sixfold miscellany of the first verse. Starting with Shakespeare and Milton, the path runs steeply downwards: the imaginary computer tells us that 'To justify the moorhens is the question', and produces the gnomic utterance 'There was below the ways that is a time.'

Lerner's mockery of what are, in effect, genetic algorithms is not entirely fair, for many potentially useful structures were generated by them. Almost every line of his poem would be intelligible in some other verbal environment. 'To justify the moorhens is the question' might even have occurred in *The Wind in the Willows*, if Ratty's friends had been accused of wrongdoing. Only one line is utter gibberish: 'There was below the ways that is a time.'

The explanation is that Lerner swapped grammatically coherent fragments, rather than single words. A similar strategy was followed (for serious reasons) by the author of *The Unfortunates*, a novel published not as a bound book but as sections loose in a box, which (except for the first and last) could be read in a random order.[11] Even Mozart (like some other eighteenth-century composers) wrote 'dice-music', in which a dozen different choices might be provided for each bar of a sixteen-bar piece.[12] In general, the plausibility of the new structures produced by this sort of exploratory transformation is increased if the swapped sections are coherent mini-sequences.

However, there is a catch – or rather, several. The first is that a self-adapting system must somehow identify the most useful 'coherent mini-sequences'. But these never function in isolation: both genes and ideas express their influence by acting in concert with many others. The second is that coherent mini-sequences are not always *sequences*. Co-adapted genes (which code for biologically related functions) tend to occur on the same chromosome, but they may be scattered over various points within it. Similarly, potentially related ideas are not always located close to each other in conceptual space. Finally, a single unit may enter more than one group: a gene can be part of different co-adaptive groups, and an idea may be relevant to several kinds of problem.

Programs based on genetic algorithms help to explain how plausible combinations of far-distant units can nevertheless happen. These inductive systems are more sophisticated than the simple sequence-shuffler imagined by Lerner. They can identify the useful parts of individual rules, even though these parts never exist in isolation. They can identify the significant interactions between rule-parts (their mutual coherence), even though the number of possible combinations is astronomical. And they can do this despite the fact that a given part may occur within several rules. Their initial IF–THEN rules are randomly generated (from task-relevant units, such as *pressure*, *increase* and *inflow*), but they can end up with self-adapted rules rivalling the expertise of human beings.

The role of natural selection is modelled by assigning a 'strength' to each rule, which is continually adjusted by the program according to its success (in controlling the pipeline, for instance). The relevant heuristic is able, over time, to identify the most useful rules, even though they act in concert with many others – including some that are useless, or even counter-productive. The strength-measure enables the rules to compete, the weak ones gradually dropping out of the system. (Whenever a new rule is generated, it replaces the currently weakest rule.) As the average strength of the rules rises, the whole system becomes better adapted to the task-environment.

The role of variation is modelled by heuristics (genetic operators) that transform the rules by swapping and inserting parts as outlined above. For instance, the 'crossover' operator swaps a randomly selected segment between each of two rules. Each segment may initially be in a rule's IF-section or its THEN-section. In other words, the crossover heuristic can change either the conditions that result in a certain action, or the action to be taken in certain conditions, or both.

One promising strategy of Plausible-Generate-and-Test would be to combine the effective components of several high-strength rules. Accordingly, the genetic operators pick only rules of relatively high strength. But the effective components must be identified (a rule may include several conditions in its IF-side and several actions in its THEN-side). The program regards a component as effective if it occurs in a large number of successful rules.

For these programs, a 'component' need not be a sequence of juxtaposed units. It may be, for instance, two sets of three (specified) neighbouring units, separated by an indefinite number of unspecified units. But the huge number of possible combinations do not have to be separately defined, nor considered in strict sequence. In effect, the system considers them all in parallel (taking into account its estimate of various probabilities in the environment concerned).

Genetic algorithms, combined with other computational ideas, might help to explain the formation of new scientific concepts – including those that are given to BACON 'for free', but which H-creative human scientists have to develop for themselves.

This possibility has been discussed by a group of authors including the initiator of genetic algorithms and the designers of the analogy-programs outlined in Chapter 7.[13] ARCS has been applied to scientific examples, and one of its co-designers has defined a constraint-satisfaction procedure for evaluating 'explanatory coherence', which can show why phlogiston-theory is a less satisfactory explanation of combustion than oxygen is.[14] (Constraint-satisfaction is involved, too, in 'revolutionary' scientific thinking: we noted in Chapter 4 that, at such times, scientists 'have to judge alternative explanations not by a single test but by many different, and partially conflicting, criteria – some of which are not even consciously recognized'.) The possibilities are exciting. As yet, however, a functioning computer-scientist that can out-Bacon BACON does not exist.

'Creative' programs rest, in large part, on hypotheses about how creativity takes place in human minds. That is, they are part of the search for a scientific psychology.

But many people believe that *no* scientific theory – whether computational or not – could possibly explain creativity. Often, their belief springs from their conviction that creative thought is essentially unpredictable. If it is, then (so their argument goes) creativity lies forever beyond the reach of science. We must now ask whether they are right.

9

Chance, Chaos, Randomness, Unpredictability

Chance, chaos, randomness, unpredictability: what do these have to do with creativity? I have argued that creative thinking is made possible by constraints, which are the opposite of randomness. Yet many people see unpredictability as the essence of creativity. How can these views be reconciled?

We must remember the distinction between psychological and historical creativity. The former is the more fundamental notion: H-creativity is a special case. Many P-creative ideas can actually be predicted. For instance, people typically ask certain exploratory questions, and notice certain structural facts, about the necklace-game. Someone's achievement on seeing that one could make 'a lo-o-o-ng necklace', or that subtraction-by-necklace would require new rules, is no less psychologically interesting because it can be foreseen. However, it is less historically interesting. All H-creative ideas are (so far as is known) *unpredicted*, since an H-creative idea is one which (again, so far as is known) no one had ever thought of before. Whether H-creative ideas are in principle *unpredictable* is another question.

In arguments about that question, the four concepts listed above often crop up. But they are all used sometimes 'for' creativity and sometimes 'against' it. For each of them supports contrary intuitions in our minds.

Chance is held to be a prime factor in many creative acts, such as Fleming's discovery of penicillin. Sometimes, however, it nips creativity in the bud. The eighteenth-century anatomist John Hunter, trying to prove that syphilis and gonorrhoea are different diseases, infected himself with pus taken from a syphilitic sailor who by chance had gonorrhoea too; Hunter went to a gruesome death believing his unorthodox view to be false.

Chaos is contrasted with creation in Genesis. Yet it is also depicted there (and elsewhere) as the fruitful precursor of creation, the seedbed from which order blossoms.

Randomness is widely seen as incompatible with creativity. If Mozart had written his dice-music by randomly choosing every note (instead of carefully constructing sets of alternative bars), the composition of minuets would have been as improbable as the writing of *Hamlet* by the legendary band of monkeys-with-typewriters in the basement of the British Museum. (Computer scientists sneeringly use the term 'British Museum algorithm' for the systematic generation and storing of every possible state.) However, randomness did play a part when the dice-music was actually played. Moreover, random genetic mutations are seen as essential for the creation of new species. And random muscular tics are used as the seeds of exciting musical improvisations, by a jazz-drummer suffering from a neurological disease.[1]

As for unpredictability, this concept is strongly linked with creativity in most people's minds. So much so that a scientific understanding of creativity is widely regarded as impossible: creative surprise, it is often said, can never be anticipated by determinist science. But unpredictability has positive associations with science, as well as negative ones. For modern science is not wholly determinist: quantum indeterminism lies at its foundations. Indeed, even strictly determined processes, whose underlying principles are known, may be unpredictable – as we shall see.

Our four key words, then, speak with double tongue: uncertainty makes originality possible in some cases, but impossible in others. To understand the tangled relations between creativity and uncertainty, we must clarify the meanings of this verbal quartet. Also, we must ask whether scientific understanding necessarily carries predictability with it. The assumption that it does underlies the anti-scientific fervour of the romantic and inspirational views. If that assumption were to fall, a scientific account of creativity might not look so impossible after all.

Sometimes, 'chance' means the same as randomness. So we speak of 'games of chance', like those played at Monte Carlo, whose outcome depends on some random factor such as the fall of a die. We even say that the British Museum monkeys, randomly tapping their typewriters, could not create *Hamlet* 'by chance'. But, in discussions about creativity, 'chance' often means not randomness so much as either serendipity or coincidence.

Serendipity is the finding of something valuable without its being specifically sought. The happy accident of Fleming's discovery is a case in point. If the dish of agar-jelly had not been left uncovered (either because its user forgot to cover it, or because the lid was accidentally knocked off), or if the window had not been open, the *penicillium* spores would never have settled on the nutrient, and Fleming would now be forgotten. Modern antibiotics owes its existence to an untidy laboratory!

Coincidence may have played a part, too. A coincidence is a co-occurrence of events having independent causal histories, where one or more of the events is improbable and their (even less probable) co-occurrence leads directly or indirectly to some other, significant event. Perhaps Fleming's usually scrupulous and previously celibate jar-coverer was in a hurry to get to an unprecedented lovers' tryst; and/or perhaps a colleague had unjammed the window to call to a long-lost friend happening to pass by. In either case, his discovery would have been due partly to coincidence.

Although serendipity is sometimes due to coincidence, they are not the same thing. For serendipity need not involve any inherently improbable event. No coincidence would have been involved in Fleming's epochal finding if his assistants had been uniformly sloppy and uniformly addicted to fresh air, and if he had been in the habit of inspecting the lab-benches and window-sills every day. Likewise, if Kekulé's general ability to transform two-dimensional shapes just happened to produce a closed curve (one of the various possibilities sketched in Chapter 4), that would have been a case of serendipity but not coincidence.

And what of Proust's eating a madeleine, which triggered the flood of memory described in *À la Recherche du Temps Perdu*? Given the popularity of these confections among French bakers, and Proust's sweet tooth, this was serendipitous rather than coincidental. So too was Coleridge's reading about 'Cublai Can', which contributed to his vision of Xanadu.

As for Coleridge's reading about the 'burnished gloss' of sea-animals, which blended into his image of the water-snakes, this may or may not have been serendipitous. He had planned for some time to write a poem about an old seaman (there is evidence that he had in mind the missing *Bounty* mutineer Fletcher Christian, who had been at school with Wordsworth). He read and re-read many passages about sea-voyages and sea-creatures accordingly – jotting some down in his notebooks, as we have seen. If he found the phrase 'burnished gloss' during this purposeful literary trawl, his finding was not serendipitous. If he came across it while reading something for a wholly unconnected reason, then it was. Since the phrase occurs in Captain Cook's memoirs, serendipity is almost certainly not the explanation of his finding.

Serendipity is made possible, for example, by computational processes like those outlined in Chapter 6. We saw there how pattern-completion and analogical pattern-matching can take place 'spontaneously', and how a subtle regularity can be noticed even though the system is not primed to look out for it. (Noticing it, of course, is not enough: checking its relevance involves analogical mapping like that outlined in Chapter 7, and sometimes experimental verification too; Kekulé's hunch about the benzene ring, for instance, needed modification as described in Chapter 4.)

But 'low-level' associative memory is not the only source of serendipity. In general, activities and skills (including those defined in terms of high-level structural constraints) that can function in parallel may interact in unplanned and unforeseen ways. By means of parallel processing of various kinds, human minds are well suited to have serendipitous ideas.

It is not always easy to decide whether a creative idea happened 'by chance', in the sense of being due to a coincidence. On the one hand, a genuine coincidence may falsely be thought to be due to some shared causal factor – as if someone superstitiously believed that the muse Terpsichore arranged for both John Lennon and Paul McCartney to go to the same school in Liverpool. On the other hand, something that we believe to be a coincidence may not really be so, if the co-occurring events actually share some crucial aspect of their causal history.

Hunter's tragic fate, for instance, was due not to coincidence but to accident, to an unexpected mishap: both the venereal diseases in question have a significantly similar causal history. (Indeed, the orthodox belief that they were a single disease had arisen precisely because most people who had contracted the one had also contracted the other as a result of similar behaviour.)

What about the near-simultaneous discovery of evolutionary theory by both Darwin and Wallace? This looks less of a coincidence when one remembers that the idea of evolution was a commonplace among mid-nineteenth-century naturalists, that its mechanism was still a live question, and that many educated persons (not only these two) would have read Thomas Malthus on the winnowing of populations through pressures on food-supply. In general, simultaneous discoveries (which, as remarked in Chapter 2, are very common) owe much less to coincidence than is often thought.

Nor is coincidence a reliably benign influence, for it can damage the creative process just as it can foster it. One of the unhappiest accidents in literary history was the surprise visit of the person from Porlock to Coleridge's cottage, without which interruption *Kubla Khan* would surely have been longer.

Coincidence is unpredictable, because we cannot foresee the improbable co-occurrence of causally independent events (in the terminology to be explained below, it is R-unpredictable). As for serendipity, there is usually no practical possibility of forecasting that something will be found without being specifically sought.

Only very occasionally can serendipitous P-creative ideas be foreseen. For instance, a parent might deliberately leave a new gadget on the dinner-table, hoping that the child will try to fathom how it works. The gadget, let us assume, was carefully chosen to illustrate some abstract principle featuring in the child's unfinished physics homework. The parent can predict with reasonable confidence that tonight's homework session will be less frustrating than yesterday's. From the child's point of view, however, its P-creation of the physical principle concerned (over dinner, not homework) was grounded in serendipity.

Both serendipity and coincidence, then, are in practice unpredictable. So the countless creative ideas that owe something to these two sources are, in some respect, unpredictable too. If science must be predictive, then the influence of chance in many cases of creativity ensures that those who seek a scientific understanding of creation will necessarily be disappointed. But if (as will be argued later) it need not, the surprise-value of serendipity and coincidence is no threat to it.

Granted that chance often plays an important role in the origin of new ideas, creativity cannot be due to chance alone. We have considered many examples in previous chapters, drawn from both art and science, which show that structural constraints and specialist knowledge are crucial. In short, Fleming was not *merely* lucky.

It was Fleming's expertise in bacteriology which enabled him to realize the significance of the clear (bacteria-free) areas surrounding the greenish colonies of mould, and which primed him to notice them in the first place. As his illustrious predecessor Louis Pasteur put it, fortune favours the prepared mind. Indeed, the words 'valuable' and 'significant' (in the definitions of serendipity and coincidence, above) imply some form of judgment on the part of the creator. Fleming was able to value the polluted dish as significant, where others would have seen the pollution as mere dirt to be discarded. Chance with judgment can give us creativity; chance alone, certainly not.

What of chaos? This word has two familiar meanings, one of which is utter confusion or disorder. Chaos in this sense is the antithesis of creativity, because it lacks the essential element of ordered judgment, in accordance with the high-level creative constraints concerned. The other meaning (the first to be listed in my dictionary) harks back to Genesis: 'the shape of matter before it was reduced to order'. In this sense, chaos – though still contrasted with creation – is seen as a precursor of it.

Whether chaos is a necessary forerunner of God's creation we may leave to the theologians. But it may be an essential precondition of some creations attributed to humans rather than gods. Indeed, 'Chaos' is the title of the first chapter in the literary study of *The Ancient Mariner* mentioned in Chapter 6.

In Livingston Lowes' words, 'the teeming chaos of the Note Book gives us the charged and electrical atmospheric background of a poet's mind'. He describes the blooming, buzzing confusion engendered by Coleridge's catholic reading and argues that every stanza was forged from this chaotic material. The water-snakes for instance (as we have seen) twist, turn and leap up from the depths not only in the Mariner's sea but also in Coleridge's mind. To be sure, passing from chaos to creation requires the formative hand of judgment, or what Coleridge called the poetic imagination. But a rampant disorder, a medley of elements drawn from widely diverse sources, can give rise to stanzas as limpid as this:

> Beyond the shadow of the ship,
> I watched the water-snakes:
> They moved in tracks of shining white,
> And when they reared, the elfish light
> Fell off in hoary flakes.

As for computational models wherein order arises out of chaos, think of the pipeline-program outlined at the end of Chapter 8. It starts off with a chaotic (randomly generated) collection of IF–THEN rules, and uses its genetic algorithms to arrive at a highly efficient set of new rules.

'Chaos' has a less familiar meaning also, in which it names a recent branch of mathematics: chaos theory. Chaos theory, whose applications range from weather-forecasting to studies of the heart-beat, describes complex systems which (at a certain level of description) are *deterministic but in practice unpredictable*. At other levels of description, chaos theory has found some previously unsuspected regularities. We shall come back to it later, when discussing unpredictability and science.

'Randomness' can mean three different things. We must distinguish these three senses, because they have different implications concerning determinism – which many people see as incompatible with creativity.

The first two meanings are very closely related. 'Absolute' randomness (A-randomness, for short) is the total absence of *any order or structure whatever* within the domain concerned, whether this be a class of events or a set of numbers. (It is notoriously difficult to define A-randomness technically, but for our purposes this intuitive definition will do.) 'Explanatory' randomness (E-randomness) is the total lack, in principle, of *any explanation or cause*.

Strictly speaking, E-randomness is the more important notion from our point of view, for our particular interest is in whether creativity can be scientifically explained. But if an event is A-random, it must be E-random too. (Since explanation is itself a kind of order, A-randomness implies E-randomness.) It follows that it is often unnecessary to distinguish between them, and I shall use the term 'A/E-randomness' to cover both.

Occasionally, however, the distinction must be made explicit. Imagine, for instance, a long series of coin-tossings which just happens to give alternate 'heads' and 'tails' throughout. Although this pattern of coin-tossings is extremely improbable, it is conceivable. There is order here, so the series is not A-random; but there is no cause, or explanation, of its structure. To be sure, each individual coin-fall has (physical) causes. But the *series* of alternating 'heads' and 'tails' does not. In short, here we have E-randomness without A-randomness. (This example shows that the level of description at which we choose to look for randomness can be crucial; we shall recall this point below, in discussing quantum physics.)

'Relative' randomness (R-randomness) is the lack of any order or structure *relevant to some specific consideration*. Poker-dice, for example, fall and tumble R-randomly *with respect to both the knowledge and the wishes of the poker-players* – as you may know only too well. They also fall R-randomly with respect to the pattern on the wallpaper, but nobody would bother to say so. In practice, R-randomness is always identified by reference to something people might have regarded as relevant (if you shut your eyes very tight and whisper, 'Six, six, six . . .', will the poker-dice oblige?). In discussions about randomness and human creativity, the potentially relevant 'something' is usually the creator's own knowledge, the structure of conceptual constraints into which the novel idea may be integrated.

If an event is A/E-random, it must also be R-random with respect to *all* considerations. But an R-random event need not be A/E-random, since it may be strictly constrained (and even predictable) in some terms *other than* the respect by reference to which it is R-random.

Poker-dice, for instance, are subject to the laws of gravity (which is why they can be 'loaded' in a crooked casino). The random firing of neurones mentioned in Chapter 6 is caused by the gradual build-up of neurotransmitter substances at the synapse. And an involuntary muscular tic might be due to identifiable chemical processes localized at the nerve-muscle junction, processes not controlled by messages from the brain and so not under the influence of the person's wishes or conscious control. So whereas A/E-randomness necessarily implies indeterminism, R-randomness does not.

Whether all three types of randomness actually occur is a controversial question. There is no disagreement about whether R-randomness happens; even determinists allow that it does. Quantum physicists hold that some events are A/E-random (as we shall see in discussing unpredictability, below). But strict determinists believe that A/E-randomness is like the unicorn: an intriguing concept that does not apply to anything in the real universe.

The various meanings of 'randomness' do not neatly divide between anti-creative and pro-creative randomness. Consider genetic mutations, for example.

Some mutations of single genes are certainly not A/E-random, for they are caused by chemical substances that affect the gene in accordance with known biochemical laws. Others may be A/E-random. If, as quantum physics implies, the emission of an individual X-ray is A/E-random, then mutations caused by X-rays are in part A/E-random too. If not, then they may be wholly deterministic.

But evolutionary biologists, who are interested in the creative potential of genetic mutations, need not care which of these is true. For their purposes, what is important is that the mutations be R-random with respect to their adaptive potential. That is, a mutation does not happen because it has survival value, but is caused in some other way – which may or may not be A/E-random.

It is R-randomness which is essential for the evolution of species. The rich diversity of biologically unconstrained mutations makes it likely that some will have survival value; natural selection can be relied upon to weed out the others. Indeed, some bacteria, if placed in a potentially lethal environment, can increase the rate of certain (non-specific) types of mutation; as a result, they may be able to use a new food-source which they could not use before. (It has recently been suggested that some bacteria can trigger mutation of the specific gene relevant to a particular environmental condition; but this is still highly controversial.)

To be sure, many changes in chromosomes (as opposed to individual genes) may not be entirely R-random with respect to survival value. Biological 'heuristics' such as crossover, as we saw in Chapter 8, provide constraints-for-adaptiveness, so that biologically plausible changes take place more often than they otherwise would. But even these processes break the chromosomes at R-random points. Only if (what is in practice impossible) these 'Lamarckian' constraints could *guarantee* the occurrence of highly adaptive transformations would R-random processes of genetic change be unnecessary for evolution.

Up to a point, similar arguments apply to human creativity. No poet, no scientist, no advertising copy-writer – and no computer program either – can be *guaranteed* always to produce an apt idea. Admittedly, some people can produce P-creative ideas much of the time, and a few – Shakespeare, Mozart – are even reliably H-creative. Such consistency cannot depend crucially on random events; it involves the disciplined exploration of highly structured conceptual spaces, as we have seen. But even Shakespeare and Mozart were presumably not averse to the 'inspiration' of accident, knowing how to exploit it better than almost anyone else.

Moreover, creative constraints (rules of metre and harmony, for example) can leave many options open at certain points in one's thinking, in which case a mental or environmental tossing of a coin is as good a way to decide as any. The distinctive style of an individual artist may depend, in part, on this. Someone may have a fairly constant, and idiosyncratic, way of deciding what to do when the general art-form – sonnets, Impressionism or dress-design – leave room for choice.

In short, human creativity often benefits from 'mental mutations'. R-random phenomena such as serendipity, coincidence and unconstrained conceptual association (what advertisers and management-consultants call 'brainstorming') are useful, because they provide unexpected ideas that can be fed into a structured creative process.

Even neurological disease can play such a role. The jazz-drummer's serendipitous tics, for instance, are almost certainly not A/E-random –

but they are R-random with respect to music. Therein lies their power, for they provide surprising rhythmic ideas which conscious thought (and Longuet-Higgins' metrical rules) could never have produced but which musical mastery can appreciate and exploit.

Mastery, involving both associative memory and deliberate judgment, is crucial. Like natural selection in biology, it enables us to take advantage of randomness, to recognize and develop its relevance. Mozart might conceivably have got a few ideas for a symphony from throwing dice, but he would have assessed their significance in musical terms. The British Museum monkeys, were their random finger-tappings ever to produce 'To be or not to be', could not even recognize it as a sentence – still less as a question.

Mastery also enlarges the mental environment, for a well-stocked associative memory provides extra opportunities for the new ideas to make connections, extra 'ecological niches' in which they may prosper. This is partly why (as we shall see in Chapter 10) experience, and the motivation to acquire it, is such an important aspect of creativity.

What is useful for creativity in minds and evolution is useful for creative computers too. A convincing computer model of creativity would need some capacity for making random associations and/or transformations. Its randomizing procedures might be A-random; for example, its instructions or associations might sometimes be chosen by reference to lists of random numbers. But they need not be: R-randomness would do. Indeed, some creative programs (such as Cohen's and Johnson-Laird's) rely on random numbers at certain points, and genetic algorithms can produce order out of chaos. Moreover, some computer models spend their 'spare' time searching for analogies in a *relatively* unconstrained way. This computerized R-randomness could exist alongside more systematic (and somewhat more reliable) 'rules' for generating useful ideas, like the inductive heuristics of the BACON-family for instance.

In principle, a creative computer could find serendipitous (R-random) ideas by systematic brute search. If the machine were fast enough, and had a big enough memory, it could exhaustively try all possible combinations of its ideas whenever it was trying to be P-creative. But the time required, given a data-base of any significant size, would of course be astronomical.

Acrobats with one arm (inconceivable to AARON) would occur to the brute-force computer eventually, and acrobats with six arms also – like a Hindu goddess. But so, too, would acrobats with a cabbage for a head and pencil-boxes for feet. 'What's wrong with that?', a historian of art might ask. 'René Magritte inspired a surrealist photograph in 1936 showing a *bourgeoise* with a cabbage for a head; and Giuseppe Archimbaldo in the sixteenth century painted human faces as assemblies of fruit and vegetables. Cabbage-heads are not beyond the bounds of creativity.' Agreed – but what about a cabbage head with pencil-box feet and a Taj Mahal rib-cage (an original idea if ever there was one)?

Our imaginary brute-search computer would need enough intelligence to realize that cabbages and carrots can appear together within an integrated artistic style, whereas cabbages, pencil-boxes and palaces cannot. (If it did manage to convince itself – and us – that the last three items can make sense within one newly developed style, fair enough.) A creative computational system must be able to situate the original idea within a conceptual space defined by intelligible constraints. In short, brute-force search must be monitored by intelligence if anything creative is to result.

The fourth member of our conceptual quartet, unpredictability, is the most important of all, because it seems – to many people – to put creativity forever beyond the reach of science.

The surprise-value of creativity is undeniable. Indeed, it is an important part of the concept, as we saw in Chapter 3. The notion that a contemporary might have predicted Beethoven's next sonata is utterly implausible. We cannot even predict the Saatchis' next advertising jingle, or (often) Grandpa's next joke. As for trying to forecast what ideas a creative person will come up with in the long term, say in three years' time, such a project would be ridiculous.

But *why*? Just what sort of unpredictability is this? And does it really destroy any hope of understanding creativity in scientific terms?

An event that is unpredictable in the strongest sense is *absolutely* unpredictable ('A-unpredictable'), unforeseeable in principle because it is subject to no laws or determining conditions whatever. In other words, A-unpredictability (like A-randomness) implies indeterminism of the most fundamental kind.

Whether there are any genuinely A-unpredictable events is disputed. According to quantum physics, there are. Quantum physics claims that some physical events, such as an electron 'jumping' from one energy-level to another, are uncaused and therefore (at that level of description) unpredictable. The electron jumps this way rather than that *for no reason at all*. In the terminology introduced above, each individual electron-jump is A/E-random (it has no order, and no explanation).

However, quantum physics also claims that *large classes* of supposedly A-random events are neither A-random nor E-random, and are in practice predictable. These large classes of sub-atomic events are not A-random, because they show order in the form of statistical regularities. They are not E-random either, because these regularities can be explained by the wave-equations of quantum physics. Moreover, these equations enable the physicist to make precise predictions about the long-term behaviour (the statistical distributions) of the relevant physical systems.

Some people argue that quantum physics must be either incomplete or mistaken, because (as Albert Einstein put it) 'God does not play at dice.' In reply, quantum physicists may admit that quantum theory could conceivably be *mistaken*, although they insist that only an irrational prejudice in favour of determinism would make anyone think so. However, they refuse to admit the possibility of its being *incomplete* in the way the determinist wants. They cite a mathematical proof that quantum theory cannot possibly be extended by adding hidden variables which obey non-statistical laws, because such extensions must alter the experimental predictions in specific ways. (Since this proof was offered, in the mid-1960s, almost all of the experimental tests have come out in favour of quantum theory.) In other words, if quantum theory is correct then the A-unpredictability of quantal events is indeed absolute.

For our purposes, it does not matter whether quantum physics is correct or not. Granted, there may be quantum effects in the brain, triggering some of the ideas that enter the mind 'at random'. If so, then A-unpredictable individual quantum jumps might conceivably contribute to creativity (as other R-random events can do). But this does not put creativity 'outside science', any more than X-rays are. In short, quantum physics illustrates one of the ways in which unpredictability (even A-unpredictability) is *not* opposed to science.

The second meaning of 'unpredictability' is more important for our discussion. An event may be unpredictable in practice, in the sense that it is unforeseeable by real human beings – and/or by other finite systems, including computers. Because this sense of the term is defined *relative to* the predictor, let us call it 'R-unpredictability'. (Naturally, any event that is A-unpredictable must be R-unpredictable too, with respect to all predictors.)

As gamblers know, there are varying degrees of R-unpredictability. To say that an event is not predictable with certainty is not to say that the chance of its happening is 'evens', or 50/50. Certain circumstances can be much more propitious for certain events than other circumstances are.

In cases where the probabilities are extremely high or extremely low, a theoretical R-unpredictability can in practice be ignored. For instance, thermodynamics tells us that a snowball could exist even in Hell. But anyone who went looking for snowballs in the Sahara would be very foolish. Such a minimal degree of uncertainty is not worth worrying about in real life. In discussing R-unpredictability, I shall have in mind only situations involving a realistic difficulty in prediction.

The reasons for R-unpredictability vary. Occasionally, they are limits of principle. (For instance, the 'indeterminacy principle' of physics states that there are pairs of related measures, such as position and momentum, whose degree of precision – in certain circumstances – cannot be increased simultaneously, because the precision of one measure must fall as that of the other rises.) Sometimes, however, what is R-unpredictable today is not R-unpredictable tomorrow.

If we discover a new scientific law, invent a more accurate measuring instrument, or build a more powerful computer to do our calculations, our ability to predict may be much improved. So it may be, also, if we become accustomed to a new artistic style (sonata-form, or atonal music). Failures in predicting are often due to ignorance (of specific details and/or general principles) and/or to complication.

Consider, for instance, what happens when you walk across a rocky beach. Suppose that some physicists knew your weight, shoe-size, the pattern stamped on the sole of your shoe, and the force exerted when you place your left foot on the ground; and suppose they also knew the mass, volume, position and surface-area of ninety-seven grains of sand heaped on a rock, and the precise contours of the rock's surface. They could not predict precisely where each, or even any, of the ninety-seven tiny objects will end up (although they could predict that they will not fly five miles into the air, or execute tango-movements across the rock).

Their problem is partly ignorance. Although they know the laws of mechanics and dynamics, which govern the movement of the sand under the pressure of your foot, they probably do not know all the relevant laws concerning the behaviour of shoe-leather on a humid summer's day. And they are ignorant of many of the initial conditions (such as the way in which your left shoe-sole has worn down since you bought the shoes). But complication is a problem too. Even if all the initial conditions were known, the interactions between the ninety-nine objects are so complicated that the computer-power needed to do the calculations would not in practice be available.

But all is not lost, because physicists can often use their scientific knowledge to predict *approximately* where moving objects will end up. They assume that the measurements made by their instruments are accurate enough to be useful, even though more precise measurements are in principle possible. And they make other simplifying assumptions (for instance, that sand-grains are round). If they were interested in the fate of individual grains of sand, or if they had to justify a detective's claim that the footmarks on the beach were yours, they could say roughly what the sand would do when you trod on it. For many purposes, roughly is good enough. So complication, while it prevents precise prediction, leaves room for useful approximation.

Our ability to approximate in this way encourages a faith in the predictive capacity of science in general: one assumes that the more scientists know about the initial conditions and the covering laws, the more likely that they can (at least) make a very good guess about what will happen. We shall see below that this faith is sometimes misplaced. Some systems are so sensitive to slight changes in initial conditions – perhaps no more than the flapping of a butterfly's wings – that even good guesses are unattainable.

The R-unpredictability of human creativity, as of sand grains on the beach, is largely due to ignorance and complication – both of which are forever inescapable.

Our ignorance of our own creativity is very great. We are not aware of all the structural constraints involved in particular domains, still less of the ways in which they can be creatively transformed. We use creative heuristics, but know very little about what they are or how they work. If we do have any sense of these matters, it is very likely tacit rather than explicit: many people can be surprised by a novel harmony, but relatively few can explicitly predict even a plagal cadence. (A computational approach, as we have seen, helps us formulate – and test – theories about the mental processes concerned.)

As for the initial conditions, the raw materials for P-creativity can occasionally be identified, or even deliberately supplied (by leaving a 'serendipitous' gadget on the dinner-table, for example). But often they cannot. And identifying all the relevant initial conditions where H-creativity is concerned is out of the question. Not even Dorothy Wordsworth knew precisely which books Coleridge had read throughout his life, or which specific passages of his current reading had interested him enough to be jotted down in his notebooks. We can never know all the contents of someone's mind that might lead to some future creative insight.

It is difficult enough to do this psychological detective-work after the fact. Had Kekulé never mentioned his phantom snakes, glimpsed gambolling in a daydream, historians of science would never have discovered them. If Coleridge had kept no notebooks, how could anyone work out retrospectively which ideas and experiences had led to his poetic creation of the water-snakes? Even Proust, whose evocative 'little tune' is subtly recalled again and again in his novel, could not identify all the tunes sounding associatively in his memory. Ignorance of the initial conditions, then, is inevitable.

The mind's complication, too, will always plague us. Complication is a problem even in computers, where all the initial conditions may be known. Locating a bug in large programs of the traditional type is in practice not always possible, and deciding why a self-organizing connectionist network passed through a particular state can be more difficult still. Prediction is even more elusive than backward-looking explanations. As for people, whether Gauguin or Grandpa, the complications multiply. The brain's conceptual networks and associative mechanisms, not to mention its exploratory strategies, are rich and flexible enough to generate infinitely many new patterns that we cannot foresee.

In some cases, including Kekulé's snakes and Coleridge's water-snakes, plausible suggestions can be made about how a creation might have originated. If what you want to know is *how it is possible* for these ideas to have played their creative roles, then a carefully argued plausibility may well satisfy you. But if you want to predict the water-snakes as you can predict tomorrow's sunrise, or even your own (unscheduled) death, you will be disappointed. In short, even if we did know the entire contents of someone's mind, the complication produced by their associative powers would prevent detailed prediction of their thoughts.

Ignorance and complication together make creativity safe for the creators. There is no hope – or no threat, if you prefer – that science will ever enable psychologists to compose all future symphonies or win all future Nobel prizes. Even Grandpa's jokes are immune to such an indignity. (Indeed, Grandpa's jokes are, very likely, less predictable than Mozart's quartets, which are richly structured by musical constraints.)

The word 'indignity' suggests that what is at stake here is our pride. We do not want to think that creativity is predictable, because we like to glory in the fact that it is not. Grandpa understandably feels miffed if someone jumps in to anticipate his joke. People who earn their living and their self-respect by continually producing H-creative ideas would feel threatened likewise, if their next symphony or scientific theory could be predicted by someone (or something) else. Some threat is experienced even by those of us who, not being H-creative ourselves, take pride in our capacity to understand and enjoy the H-creations of others. For that capacity is an aspect of our own P-creativity (as previous chapters have shown).

This self-regarding attitude is one of the reasons for the widespread resistance to determinist accounts of creativity. Of course, even the most committed determinist does not claim that full prediction will ever be possible in practice. The claim is that, because there is no A/E-randomness or A-unpredictability, creative ideas (like everything else) are predictable in principle. However, the determinist may well believe that *some* creative ideas that are currently R-unpredictable will be successfully predicted in the future.

Is that possible? Or would an idea's being predicted show that it wasn't really creative after all?

If a psychologist – or a computer program – were actually able to predict a composer's next symphony, the music would be no less beautiful. Moreover, the composer's achievement would be no less P-creative. It would still involve the exploration and/or transformation of the conceptual space within the composer's mind. The H-creativity, of course, would belong to the predictor. But we saw in Chapter 3 that H-creativity is not a *psychological* category. To understand creativity as a psychological phenomenon, P-creativity is the crucial notion. This mental ability is not destroyed by prediction, still less by in-principle predictability. In short, determinism is compatible with creativity.

But perhaps the unpredictability of human thought has other grounds, too? Even supposing determinism to be true, could the unexpectedness of Beethoven's next sonata or Dior's next design be partly due to psychological complexities deeper than mere complication? And could these complexities lead to sudden wholesale change in the mental landscape, as opposed to a continuous associative journey through it? If so, would we finally have identified an aspect of creativity which puts it beyond the cold hand of science?

Until very recently, scientists – and most other people, too –assumed that deterministic systems can hold no such surprises. They assumed not only that R-unpredictability (if not based in a genuine A-unpredictability) could in principle always be overcome, but also that an extra decimal place in the measurement of initial conditions was always handy but never crucial. Extra precision, they thought, provides closer approximation rather than grounds for astonishment.

In other words, they shared an intuition (which turned out to be mistaken) that the mathematics needed to describe the natural world involves only functions having the sort of smoothness that allows approximation of this kind.

In many cases, of course, this 'no surprises' assumption is correct. The outstanding example is orbital motion in space. Several years ahead of time, space-scientists predicted the rendezvous between the Voyager satellite and the remote planets of Jupiter and Neptune with remarkable precision. The question is whether the assumption is true for all deterministic systems.

Consider the weather, for instance. Long-term forecasting is extremely unreliable. Even short-term forecasts can be wildly wrong: the BBC weather-man Michael Fish achieved unwanted notoriety in October 1987 by declaring, only a few hours before the south of England was devastated by hurricane-force winds, 'There will be no hurricane.' It is this sort of thing which makes people say that meteorology is not a real science.

Meteorologists, however, insist that it is. For, drawing on theoretical physics, they have identified general principles governing wind and weather – principles that can be precisely expressed as mathematical equations.

The equations concerned are differential equations. That is, they relate variation in some measures to variation in others. For example, they describe how changes in pressure affect changes in density, and *vice versa*. Meteorologists use them, together with data about air-pressure and the like, to compute how weather conditions change from moment to moment. An equation's results at one moment are fed back as its input-values at the next, and the process is repeated indefinitely. (Similarly, differential equations describe the continuous changes of the weights in the self-equilibrating connectionist networks described in Chapter 6.)

Naturally, if the measurements of weather conditions that are used at the start of this repetitive calculation are inaccurate, the later results – the weather predictions – will be inaccurate too. So much has long been agreed.

In the past, the common assumption was that improved information (from increasingly accurate instruments on weather-balloons, research-ships and satellites) would result in long-range forecasts approximating more and more closely to reality. Recently, however, meteorologists have argued that reliable long-range forecasting will forever remain impossible. We can never be confident of knowing 'enough' about the initial conditions, because the system's sensitivity to initial conditions is far greater than was previously supposed. The flapping of a butterfly's wings in Fiji could conceivably cause a cyclone in Kansas.

This view is justified by the branch of mathematics called 'chaos theory' (which, as the word 'theory' suggests, has found new regularities too). But some of its first intimations within meteorology arose from the chance discovery in 1961 that, even using relatively simple equations, a few extra decimal places can make a surprisingly large difference.[2]

A physicist modelling weather-systems on his computer had intended to repeat some calculations, starting from a point which the computer had reached in the middle of its run on the previous day. The first day's printout gave figures representing the conditions at that point, so he keyed these figures in as the starting-point for the second day.

On letting the computer run his equations on this datas, he was amazed to find that after a few cycles of repetition the resulting curve began to diverge significantly from yesterday's. Quite soon, the two curves were utterly different. So far were they from being approximations of each other that he could never have predicted the second on the basis of the first. If someone else had shown them to him, he would never have suspected that they were generated by identical equations. (The unexpected outcome was not due to any indeterminism, or A-unpredictability: although some differential equations involve a randomizing factor, these did not.)

What had happened? It turned out that the calculations actually done by the computer (and stored in its memory) were always correct to six decimal places. But in the printouts, to save space, the last three decimal places were ignored. So, on the first day, '.506127' in the memory had appeared as '.506' in the printout. When this three-place number was keyed in on the second day, the computer stored it as '.506000'. The difference, fractionally over one ten-thousandth part

(0.000127), would very often be negligible. But the two wildly diverging curves showed that in this case it was not.

This mathematical exercise proved that *some* fully deterministic systems are so sensitive to tiny variations in the initial conditions that they are R-unpredictable in a special way. Let us call them 'butterfly-unpredictable', or *B-unpredictable.* A system is B-unpredictable if adding just one more decimal point to our measurements would sometimes lead to a very different outcome, which can be computed *only by actually working through the consequences of the equations.*

The consequences have actually to be computed (as opposed to being predicted, or estimated, by means of general principles) because, where B-unpredictability is concerned, one cannot say in advance that the differences due to small variations in input will lie within certain bounds. One cannot even say that the resulting differences will always get smaller as the input variations get smaller. To put it another way, butterfly-flappings are not fed into some hugely powerful (but constant) amplifier: they do not always cause cyclones. If they did, then meteorological predictions would be possible, provided that one kept a close eye on all the world's butterflies. Rather, the relation between flapping butterfly-wings and long-term consequences is highly variable, and enormously sensitive to initial conditions.

In the jargon of chaos theory, B-unpredictable systems are called 'complex'. This is a technical term: not all chaotic systems are complex in the everyday sense, as the weather is. For instance, a pendulum mounted on another pendulum would normally be regarded as a very simple system, but under certain initial conditions its behaviour is B-unpredictable, or 'chaotic'. Similarly, conceptual spaces defined by surprisingly simple equations (dreamed up by mathematicians, not culled from physics) can show chaotic variability: change the input-number you start with by only a tiny fraction, and the equations may give you a shockingly different result. In short, exploring an apparently simple space leads one repeatedly to surprises.

Someone might object: 'Chaotic systems are predictable in principle, just like any other deterministic system, so there is no need to distinguish between B-unpredictability and other sorts of R-unpredictability'.

But this remark obscures the practical realities, for if the fourth decimal place (or perhaps the hundredth?) can suddenly make a highly significant difference then scientists are in a new situation. They cannot even say *approximately* what the system will do. (So chaotic complexity is not the same as complication, defined above.) If the antics of a Fijian butterfly could cause devastation in Kansas, long-range weather forecasting is achievable only by God. Human scientists are stuck with B-unpredictability, as well as with R-unpredictability of more familiar kinds.

You might think that scientists are not just stuck with B-unpredictability but defeated by it, that chaos theory sounds the death-knell of meteorology as a science. But this would be to notice the 'chaos' and ignore the 'theory'.

Some very surprising deeper regularities have been discovered in the behaviour of B-unpredictable systems. These new mathematical structures can be used to make high-level predictions of a kind that could not be made before. Being highly general, they are not confined to clouds and cyclones. Chaos theory is being applied also in fluid dynamics, aeronautical engineering, population biology, embryology, economics, studies of the heart-beat – and, by the time you read this, doubtless other areas too. In scientific contexts, unlike Genesis, 'chaotic regularity' is not a contradiction in terms.

One intriguing example of chaotic regularity is 'period doubling', in which a tiny increase in one condition leads repeatedly to fundamental changes in the system as a whole. A stretch of B-unpredictable behaviour suddenly gives way to a regular pattern of oscillation, which (as the increase continues) abruptly lapses into chaos again, followed later still by a further wholesale restructuring giving a new pattern of oscillation . . . and so on. The new oscillation pattern is a simple function of the pattern at the previous level (the 'length' of the oscillation-waves is halved at each stage). Mathematical chaos theory proves that, in principle, period-doubling could go on to infinity, smaller-scale patterns being originated for ever.

Nor is this a mere mathematical curiosity. Period-doubling has been shown (by computer calculations) to be a consequence of a variety of scientific theories. And it has even been observed in experimental laboratories (studying fluid flow, for instance) and in clinical cardiology.

Another mathematical concept of chaos theory is the 'strange attractor'. This marks the fact that a system's behaviour can draw closer and closer to some ideal pattern, without ever quite reaching it and without ever repeating itself. Every cycle is generated by the same equations. Yet each one is new, a close approximation to its predecessor and its successor, but never precisely the same.

The question obviously arises whether chaos theory applies to the brain. If it does, and if the brain-processes in question are involved at some stage of creative thinking, then certain aspects of creativity may be B-unpredictable. This would provide a further justification for the intuition that individual creative ideas cannot, in practice, be foreseen. However, it would also suggest the possibility of *unsuspected sorts of regularity* within creative processes.

As yet, we have only the flimsiest of evidence with which to address these questions. The possibility of chaotic effects at the neurological level cannot be ruled out *a priori*. Many (though not all) systems showing chaotic complexity are highly complex in the everyday sense of the word too – as is the brain. Moreover, chaotic systems in general are devices in which one moment's output is fed in as the next moment's input – and the neural networks in the cerebral cortex include many feed-forward circuits and feedback loops. However, the specific relevance, if any, of chaos theory to neuroscience is still an open research question, on which very little work has been done.

Some neurophysiologists have suggested that chaotic neural activity in the brain helps prevent cell-networks from being trapped into certain patterns of activity, and so underlies (for example) our ability to learn and recall sensory patterns, such as sounds and smells.[3] In other words, neuronal chaos may act rather like a randomizing device, 'shaking up' the cell-networks from time to time so that they learn a wider range of sensory patterns than they would do otherwise. Whether this suggestion is correct is still unclear, and not all neuroscientists accept it. (As we saw in Chapter 6, with reference to Boltzmann machines, neuroscientists need not use chaos theory to argue that neural networks may involve a randomizing process.)

What about the possible relevance of chaotic brain-processes (supposing that they exist at all) to creative thinking?

If there are 'strange attractors' in the brain, patterns of activation to which cells (or groups of cells) successively approximate without passing through the same state twice, then new ideas might be continuously generated accordingly. However, it is doubtful whether this effect could contribute usefully to creativity. If each idea were only marginally different from the one before, the process would not be random enough to be useful. A creative idea must be not only new, but surprising.

Period-doubling might be more relevant. A sudden restructuring of an associative field, caused by some small change, might lead to ideas that are novel in a deeper sense, ideas which could not have been produced (or even approximated) within the preceding cyles of system-behaviour. Perhaps something of this sort underlies introspective reports like Goethe's, who said that – having had the romance *Young Werther's Suffering* in his mind for two years without its taking form – he was told of a friend's suicide and 'at the instant, the plan of *Werther* was found; the whole shot together from all directions, and became a solid mass, as the water in a vase, which is just at the freezing point, is changed by the slightest concussion into ice'.[4]

As the 'mights' and 'perhapses' scattered in the preceding paragraphs imply, these suggestions are highly speculative. But suppose there is some truth in them: what would follow?

We would have one more concept (besides serendipity, coincidence, randomness and R-unpredictability) accounting for the fact that new ideas often pop up in apparently unprincipled and idiosyncratic ways. But we could not explain creativity in terms of the B-unpredictability of mathematical chaos, for precisely the same reason that we cannot explain it in terms of common-or-garden chaos.

Unpredictable new ideas can be useful only in the context of stable high-level generative principles, defining the particular conceptual space involved and guiding the exploration of it. Think of a composer committed to tonal harmony, a chemist puzzling over valencies and molecular structure, or a draughtsman drawing the arm of an acrobat pointing straight at the viewer. All these people are bound by identifiable constraints, as we have seen.

Even creative transformations of generative principles are themselves constrained, for they must yield a new (and valuable) conceptual space whose structure is recognizably related to the old one. Cabbages are acceptable in place of human heads only if they provide the audience with a way of finding their bearings in a new aesthetic space (such as surrealism).

On these matters, chaos theory has nothing to say. The relative stability of the high-level creative principles concerned is at odds with B-unpredictability. It is not even clear whether, as tentatively suggested above, some of the deeper regularities in chaotic behaviour could be involved in some creative transformations. For sure, they cannot underlie all of them. A composer's deliberate exploration of harmonic constraints, for instance, demands disciplined self-reflection, not the intermittent spontaneous restructuring of a chaotic system.

Above all, chaos theory cannot 'rescue' creativity from science. Chaos theory is increasing our scientific understanding, not destroying it. To be sure, it has changed some of our fundamental ideas about what science is like. The clockwork universe imagined in the late eighteenth century by Laplace, who believed that if he knew the position and momentum of every particle in the universe he could predict its entire future history, was already long gone by the 1960s. Now we have to recognize not only quantum indeterminacy but chaotic complexity too.

If the psychology of creativity turns out to be infested with butterflies, it will be even more difficult than we had thought. But so is fluid dynamics: creativity would be in good scientific company.

One last point must be made about unpredictability. Many people assume that prediction is the core concern of science. This is why the negative associations between 'unpredictability' and 'science' are so strong. But science is not prophecy. Its prime focus is on structured possibilities, not on facts – and certainly not on future facts as opposed to past facts. Its main aim is not to say what will happen, but to explain *how it is possible* for things to happen as they do.

A 'side-effect' of much scientific explanation is to enable us to predict, and sometimes even to control, part of what will happen. This is very useful, since all technological applications of science, and all experimental methods, depend on it. But prediction is not essential to scientific theories. Darwin did not even attempt to predict what new species would arise in the future. Rather, his theory explained how it was possible for evolution to happen at all (many of his contemporaries suspected that evolution had happened, but they could not imagine how).

A further benefit of (some) scientific understanding is the ability to explain, if we are interested, why a particular event did happen. But a doctor who can explain the origin of a patient's kidney-stone could not necessarily have predicted it. Similarly, a literary critic who can explain Coleridge's water-snakes could not have predicted them on the basis of his notebooks.

In sum, science is riddled with uncertainty: R-unpredictability, B-unpredictability and (according to quantum physics) A-unpredictability too. But this need not discourage the scientifically minded, because prediction is not what science is really about. Its task is to demystify the snakes and the water-snakes, not to predict them. Anyone hoping to understand *how creativity is possible at all* can cheerfully allow that human creations (and computerized creations too) will always be full of surprises.

10

Elite or Everyman?

'But Mozart was different!' Indeed he was. And winning a million pounds on the football pools is different from winning the price of a hot dinner.

The life a fortune makes possible is not just more-of-the-same (hot dinners *every* day), but fundamentally different: in ambition, variety and freedom. The pools-winner must develop new skills, and venture into unfamiliar conceptual territory. It will take some effort to learn to sail a yacht, or to appreciate the Picasso newly hung on the living-room wall. Even the opportunity to give large sums to charity brings its own complications, new problems that could not have arisen before. But the pools-winner needs no special faculty: native wit can do the trick.

Does the difference between Mozart and the rest of us lie in some supernatural influence or special romantic power? Or is it more like the difference between winning a fortune and winning a meal-ticket, one of which engenders a space of opportunities and problems which the other does not? Specifically, could Mozart's genius have been due to his exceptionally skilful use of a computational resource we all share: the human mind?

The romantic and inspirational 'theories' see the *historically* creative as set aside from normal humanity: H-creative insights, and H-creative people, are supposed to be fundamentally different. Intuition, not to mention divine assistance, is said to be a special power that enables H-creators to come up with their ideas. (Mere P-creators, who have ideas that are not historically new but which they themselves could not have had before, are rarely considered.)

Both these myth-like approaches claim support from people's introspective accounts of their own H-creativity. We saw in Chapter 2 that great artists and scientists have frequently reported the sudden appearance of H-creative ideas in their minds. However, what is reported as sudden may not have been sudden at all. And what seems to have no conscious explanation may involve more consciousness than one thinks.

Consider the familiar phenomenon of noticing something, for instance. Noticing and noticing how you notice are two very different things. Think of the last time, or look out for the next time, that you notice something, and try to detail as many of the possibly relevant facts about your own mind (both conscious and unconscious) that you can think of. You may find this very difficult. If so, your humdrum achievement of noticing may come to seem rather mysterious – almost as mysterious as 'insight'.

Likewise, if you try to say just what it was which reminded you of something, or to detail the missing links between the first and last ideas, you will not always be able to do so. (A computational model of reminding, using concepts like those in the BORIS program, has outlined some of the schema-relating processes that may be involved.)[1]

Your difficulty is only partly due to the hiddenness of the unconscious influences at work. For conscious thoughts can be elusive too, and people's sincere reports of them are not always reliable. Try, for instance, to recount all the thoughts that fleet through your mind while making up the next line(s) of a limerick beginning: *There was a young lady of Brighton . . .'*.

You will probably come up with a sparse harvest.

It is not easy to catch one's thinking on the wing, and detail every fleeting image. (Normally, of course, one does not even try.) This is one reason why the notion of intuition, or inspiration, is so compelling. People rarely try to capture the details of their conscious thinking, and when they do so they do not necessarily make a good job of it.

Their lack of introspective success is partly due to lack of practice: they have not learnt how to introspect in a way likely to yield rich results. If you merely ask someone to 'think aloud', you may not get very much of interest. But if you tell them how to go about it, that may help. In his fascinating discussion of creativity, the psychologist Perkins suggests six 'principles' of introspection:[2]

1 Say whatever's on your mind. Don't hold back hunches, guesses, wild ideas, images, intentions. [Notice that this is also very good advice on 'brainstorming', or 'lateral thinking'.]
2 Speak as continuously as possible. Say something at least once every five seconds, even if only 'I'm drawing a blank.'

3 Speak audibly. Watch out for your voice dropping as you become involved.
4 Speak as telegraphically as you please. Don't worry about complete sentences and eloquence.
5 Don't overexplain or justify. Analyse no more than you would normally.
6 Don't elaborate past events. Get into the pattern of saying what you're thinking now, not of thinking for a while and then describing your thoughts.

Try it! (This time, you can complete: *There was a young man of Tralee . . .*'.) You will very likely find (especially if you do this sort of thing several times) that you report a lot more going in your mind than you did in the previous introspective exercise.

In short, despite the importance of unconscious processes, myriad fleeting conscious thoughts are involved too. The fact that they are rarely reported is not decisive.

Unreliability in introspective reports of creative thinking has another cause, too. Self-reports are informed by the person's tacit theories, or prejudices, about the role of 'intuition' in creativity.

Introspection is looking into one's own mind, and it shares an important feature with looking into anything else: to a large extent, you see what you expect to see. A doctor in the midst of a chicken-pox epidemic, faced with a case of small-pox, is very likely to misdiagnose the disease. Indeed, much more startling examples of prejudice-driven misperception occur.

In one experiment, medical students were shown a photograph of a baby in a white gown with a simple frill at the neck, leaning against a brick wall. The students offered a number of diagnoses. They commented that the baby was sleeping peacefully, so certain illnesses could be ruled out. They argued about the apparent negligence involved in sitting a baby up against unyielding bricks, in contrast with the apparent care suggested by the spotless frilly nightdress. None perceived the situation rightly.

In fact, the baby was neither ill nor asleep, but dead. The nightdress was not a nightdress, but a hospital shroud (which the medical students had seen many times). And the brick wall was the wall of the hospital mortuary, which – again – they had seen on numerous occasions. Their tacit assumption that they would be shown a living baby, not a dead one, led them to misperceive the situation to such an extent that even familiar things were misinterpreted (*despite* the apparent anomalies in the situation).

If this sort of thing can happen when several highly intelligent people take twenty minutes to discuss a photograph staring them in the face, how much more likely that a fleeting self-perception may be contaminated by preconceptions about what one will – or will not – find.

If you already believe that 'insights' come suddenly, unheralded by previous consciousness, then in your own introspective experience they are likely to appear to do so. And if you already believe that they are caused by some unconscious (and semi-magical?) process of 'intuition', you will not be looking as hard as you might for causes potentially open to consciousness. (Likewise, someone trying to explain someone else's thought-processes will look rather harder if convinced that there is something 'concrete' there to find. Livingston Lowes burrowed so meticulously through Coleridge's library, with the notebooks as his guide, precisely because he *did not* believe that Coleridge's poetry was generated by unnatural means.)

Moreover, someone who is (or aims to be) regarded as H-creative, and who accepts the romantic notion that H-creative individuals are somehow set apart from the rest of us, might not wish to find too much conscious richness in their mind. And what one does not want to find, one does not assiduously seek.

For all these reasons, then, introspective accounts of creative episodes cannot be taken at face value. Even if (which may not be the case) they are full and accurate reports of the person's conscious experience, they are structured by preconceptions much as 'outer' perception is.

Similar caveats apply to memory. Psychological experiments have shown that people's memories of specific events depend very largely on their general assumptions, on the conceptual structures that organize their minds. Broadly, only items that fit into the conceptual spaces within one's mind can be stored there: items that do not fit are 'squeezed' into (or rather, out of) shape until they do.

This casts further doubt on the reliability of 'introspective' accounts of creative insight, since most of these are not introspections but retrospections. Artists and scientists are usually far too interested in what they are creating to be bothered, at the time, to focus on how they are creating it. Moreover, the importance of the idea is often not fully realized until long afterwards. Only then does the creator, perhaps egged on by an admiring public, set down the (supposed) details of what actually happened in his or her consciousness at the time. All the more opportunity, then, for the creator's preconceived ideas about the creative process (and the 'specialness' of creators?) to affect the description of what went on.

For example, Coleridge's well-known account of how he came to compose *Kubla Khan* (which he subtitled 'A Vision in a Dream') conflicts both with other self-descriptions of this episode and with documentary evidence. The best-known account comes from his Preface to the poem, published in 1816. There, Coleridge says that in 1797 (a full twenty years earlier) he 'fell asleep', and remained for some hours in 'a profound sleep, at least of the external senses'. But in 1934 a manuscript in Coleridge's handwriting was discovered which gave a slightly different version of the poem, and which referred to 'a sort of Reverie' rather than a 'dream' or 'sleep'. Internal evidence suggests that this (undated) version was written earlier than the poem as he published it, for several of the ways in which it differs from the familiar version are closer to the sources, such as *Purchas's Pilgrimage*, which are known to have influenced his composition.

In discussing this case, Perkins points out that besides being a prisoner of his own memory and his own theories of the creative process (as all of us are), Coleridge was not over-scrupulous about getting his checkable facts right. His own contemporaries regarded him as untrustworthy on dates of composition (the discrepancy was sometimes considerably more than a year or so). And literary historians have detailed a number of examples where his 'factual' reports simply cannot be accepted.

It does not follow that Coleridge was a rogue, or a fool either. Simply, he was human, and subject to the limitations of human memory (and to the temptations of laziness). The point here is a general one, applying to others besides Coleridge. Descriptions written years after the event are interesting, and may be used as evidence. But they cannot be taken as gospel.

Something else that cannot be taken as gospel is Poincaré's view (widely accepted by writers on creativity) that incubation – time away from the problem – involves a special sort of extended unconscious thinking. To be sure, his insistence that it involves more than just a refreshing rest appears to be correct (Perkins has done several experiments to this effect). But there are other possible explanations why a change of activity may be followed by a creative insight that surmounts the original difficulty.

For instance, one's mind may turn to an absorbing problem at many times during the day, perhaps while brushing one's hair or doing the washing-up. As Perkins puts it, 'time away from the desk' is not necessarily 'time away from the problem'. – Or one may be just on the point of solving some problem when an interruption occurs. On returning to the problem some time later, the solution that was about to pop up at the earlier moment may emerge now. The explanation here is memory, rather than any 'incubatory' thinking. – Alternatively, one may have picked up some cue, either consciously or unconsciously, during one's time away from the problem. This is serendipity (noticing), not 'incubation'.

Again, sleep provides time away from conscious thinking about the problem. It also seems to allow some relaxation of the logical constraints which are respected in the waking state (hence the many reports of original ideas occurring to someone as they wake).

Finally, the feeling that such-and-such an approach, on which one has already invested a great deal of effort, *must* be the right way ahead can block a solution. But this feeling may be weakened if, for a while, one thinks about other things and stops worrying about how to solve the problem. It is understandable, then, that shifting attention to a different (perhaps equally difficult) problem sometimes helps a person to master the first one.

As Perkins remarks, none of this proves that a special sort of incubatory thinking never happens. But there is no firm evidence that it does. And there is plenty of evidence in support of several alternative explanations of why leaving a problem for a while can often be helpful. In sum, there is no reason to believe that creativity involves unconscious thinking of a kind utterly different from what goes on in ordinary thought.

Creativity draws crucially on our ordinary abilities. Noticing, remembering, seeing, speaking, hearing, understanding language and recognizing analogies: all these talents of Everyman are important. So is our ability to redescribe our existing procedural skills on successive representational levels, so that we can transform them in various ways. It is this which enables young children to draw increasingly imaginative 'funny houses' and 'funny men', as we have seen – and one could hardly get any more ordinary than that.

To say that something is ordinary, however, is not to say that it is simple. Consider Kekulé's vision of snakes, for instance.

When we discussed Kekulé's reports of his experiences by the fireside, and on the omnibus, we took a great deal for granted. We asked how he managed to come up with the analogy between snakes and molecules, but we did not question his ability to be reminded of snakes in the first place. We took it for granted that Kekulé – like the rest of us – could see snakes as having certain spatial forms, and that he could notice that one had seized hold of 'its own tail'. We took for granted, too, his ability to distinguish 'groups' of atoms, and to identify 'long rows', or 'chains'. We simply assumed that he could see some atoms as 'smaller' and others as 'larger'. And we raised no questions about his seeing snakes 'twining and twisting', or atoms 'in motion'.

How are these achievements possible? For achievements they are, achievements *of the mind* (A camera can do none of these things.) Their mind-dependence is not due to the fact that Kekulé's snakes, and his gambolling atoms, happened to be imaginary. Comparable questions arise with respect to seeing real snakes. Suppose Kekulé had noticed a tail-biting snake while strolling in the countryside. We could still ask, for example, how he identified the tail as the snake's 'own' tail. Much as similarity is a construction of the mind, so is visual form.

The perception of visual form seems, to introspection, to be simple and immediate. Apparently, we 'just see' snakes, snails and snowmen. But these introspections are misleading, for even such everyday seeing is not psychologically simple. On the contrary, it requires some fancy computational footwork.

To see snakes, and to imagine them, Kekulé had to be able to recognize individual figures, as distinct from the background. He had to pick out both spots and lines, and to do the latter he had to identify both continuity and end-points. (A snake biting its own tail is all continuity and no end-points.) He had to appreciate juxtaposition and distance, if he was to see 'groups' of atoms, long rows 'more closely fitted together', or a larger atom 'embracing' two smaller ones. And to see a row as 'long', or an atom as 'larger' or 'smaller', he had to judge relative size.

The interpretative processes involved are neither obvious nor simple. Even finding the 'lines' is difficult. Consider a photograph (or a retinal image) of someone wearing a black-and-white striped tie. You may think that identifying the stripe-edges is easy. 'Surely,' you may say, 'each edge is a continuous series of points at which the light-intensity changes sharply: bright on one side, dark on the other. All one needs, then, is a tiny physical light-metre that can crawl over the image and find those points.'

Well, yes and no. Physical devices to pick up sudden changes in light-intensity are indeed needed, and they exist both in the eye and in many computers. The problem is that, in most real situations, there is *no* continuous series of change-points in the image which exactly matches what we perceive as a line. In general, the edge of a physical object (such as a snake), or of a marking on a physical surface (such as a stripe), *does not* correspond to any clear, continuous series of light-intensity changes in the physical image reflected off it. In the image, there will be little segments of continuous intensity-change – but there will be gaps and offshoots as well. What is needed is a device that can recognize that the *significant* segments are colinear across the gaps, whereas the offshoots are not. What is required is not just physics, but also computation.

If the viewer is looking at a dalmatian dog lying on a zebra-skin rug, the line-finding device may require some help from depth-detectors. A single black region in the image may represent *a black dog-patch adjacent to a black zebra-stripe*. With respect to physical light-intensities, there may be no distinction here between the doggy part of the image-region and the ruggy part. Mere colinearity (with the lines representing the adjoining contours of the dog's back) may not settle the matter, if there happen to be similar problems with regard to other dog-and-rug regions in the image.

But depth-detectors can help. The images falling on the left and right eye differ slightly, according to the object's distance from the eyes. Consequently, a systematic comparison of corresponding point-images can detect depth-contours (where one physical surface lies some way *in front* of another). By this means, then, the visual system can find the contour of the dog's body within the uniformly black region of the image. And if that depth-contour is colinear with one of the many line-segments running into the black region, then that line-segment is probably the one which represents the dog's back.

There are texture-detectors in the visual system too, which can compute texture-differences between adjacent parts of the image. In the case of the furry dog lying on the furry rug, these might not be of much help. But if the dalmatian were lying on a black and white lino floor, they could help to disambiguate the dog-and-lino image-regions. (Such multiple constraint-satisfaction can be effected by parallel processing, as we saw in Chapter 6.)

If Kekulé had seen a viper lying on stripey grass and twigs, he would have needed not only line-detectors, but depth-detectors and texture-detectors too. Motion-detectors would also help to make the snake visible, since image-lines that move together normally represent real object-edges. (This is why many animals 'freeze' when they sense predators.) In short, to see snakes twining and twisting, or biting their own tails, requires complex computational processes.

Everyday visual interpretation is relevant not only to seeing snakes and imagining benzene-rings, but to the visual arts too. Think, for example, of the intricate line-drawings of John Tenniel, or the (much simpler) acrobats sketched by AARON. How are the individual lines identified, even in smudged newspaper reproductions? And how are they interpreted, as the hem of Alice's dress or the bulging biceps of an acrobat's arm?

Again, remember the Impressionist movement in the late nineteenth century. Or consider Picasso's creative progression from the relatively realistic paintings of his charming Blue and Rose periods, through the austerely analytic Cubism of *Les Demoiselles d'Avignon* and *Girl with a Mandolin*, to the distorted 1930s portraits of his mistress Dora Maar showing her with two eyes on the same side of her nose. Many art-connoisseurs at the time scorned these new styles as unnatural, unreasonable and (therefore?) ugly. Some people still do. But a computational psychology can help us to understand something that was intuitively grasped by the artists concerned (sometimes backed up by reference to the scientific theory of the time). Namely, such painting-styles are grounded in the deep structures of *natural* vision.

There are no natural situations (Narcissus' pond excepted) in which we can see from two viewpoints simultaneously. Because our eyes are in the front of our heads, and we cannot be in two places at once nor assume two bodily attitudes at the same time, we always see things from a single viewpoint. This fact is deliberately exploited in computer models of vision, whose interpretative heuristics work only because it is true. The biological visual system, in effect, takes it for granted too: our natural visual computations assume a single viewpoint.

No wonder, then, if we experience a shock of surprise on seeing Dora with (apparently) two eyes on the side of her face. Such a thing has never been seen before – and it *could* never be seen, in the real world. Our visual machine simply does not permit it.

But who ever said that the artist must accept all the constraints of the real world? Enough if he can use them, challenge them, transform them, in ways that are somehow intelligible to us. The pictures of Dora are intelligible (they are even recognizable, if one has seen a photograph of Dora). She does have two eyes, after all; and she does have a nose with a Roman profile. Simply, we cannot in real life see her as having these three things together. If the painter chooses nevertheless to depict all three on the one canvas, why should we complain? Is he really doing something *utterly* unnatural, with no intelligible grounding in our knowledge and visual experience? Or is he, rather, exploring the conceptual space within which things may be seen either frontally or in profile?

Similarly, if the Cubist chooses to analyse visual form in geometrical terms, what is wrong with that? Why should Picasso have had to keep his *Demoiselles* rolled up in his studio for twenty years, spurned even by his closest admirers and friends? What was wrong with Cézanne's advice to a fellow-artist to 'deal with nature by means of the cylinder, the sphere, and the cone'? It is a valid aesthetic question, how far a painter can intelligibly depict nature in these ways.

(Even scientists may approach nature by using similar ideas. Some psychologists have tried to explain our perception of spatial forms in terms of 'generalized cylinders'. The idea is that the visual system computes the shape of a wine-bottle as a long, narrow cylinder whose diameter is especially narrow at the top; a sugarlump would be a short, fat cylinder with a squared cross-section; and a snake would be a very long, very thin, cylinder with a curved axis. This method of representation is used for some special applications, but is not widely accepted. How, for instance, could it capture the shape of a crumpled-up piece of paper?)

Impressionists focused on patches of light, rather than realistic visual interpretations. A painter like Monet can help us to realize that distinguishing colour-patches is one thing, and seeing them as water-lilies is another. Indeed, computational theories (and computer models) of vision suggest that our visual perceptions are constructed on several successive representational levels.[3] Colour-patches and line-segments are identified at a relatively early stage. Physical surfaces, located relative to the current position of the viewer, come later. Solid objects, independently located in three-dimensional space, come later still. And named things, such as water-lilies, are constructed last of all. What the Impressionists did, in effect, was to remind us of (some of) this, and to suggest what our vision would be like if we could not compute interpretations at the higher levels.

The Impressionists were well aware that their artistic style is relevant to visual psychology, which they discussed at some length. Other painters, too, have been influenced by scientific theories. Bridget Riley's canvases, for instance, are based on psychological studies of visual illusions. The pointilliste Seurat, who chose his palette by reference to theories of optics, even wrote to a friend: 'They see poetry in what I have done. No, I apply my method, and that is all there is to it.'[4]

But most creative artists are content to ignore theoretical questions about how the mind works. They take our everyday abilities for granted, even while tacitly exploiting their subtleties in their work.

John Masefield did not need a course in phonetics or speech-perception to contrast the mellifluous 'Quinquireme of Nineveh' so effectively with the 'Dirty British coaster with a salt-caked smoke-stack'. Nor did the director of the James Bond film *Dr No* need a degree in psychology to know that British cinema-goers would notice Sean Connery noticing the portrait of the Duke of Wellington, recently stolen from the National Gallery, in Dr No's lair. Private, and not-so-private, jokes like this one are legion in the arts: think of the allusions in *The Waste Land*. (Even nuclear physicists occasionally play such games; why else would they speak of 'quarks'?) Such delights are possible because artists have a good intuitive grasp of what the human mind can do.

The psychologist, however, cannot take our ordinary abilities for granted. Rather, the aim is to understand them as explicitly as possible. How do we manage to notice something? How do we remember things, how do we understand English sentences, and how do we appreciate analogies? A computational psychology can help us to identify the detailed mechanisms that underlie everyday capacities.

Without these mechanisms, creativity (and its appreciation) would be impossible. No noticing, no Newton. No analogy, no Antonioni. And for sure: no memory, no Mozart.

That Mozart had an exceptional memory, at least for music, is clear. Anecdotes abound, for example, about his ability to write down entire cantatas after having heard them only once (and to imagine whole symphonies before hearing them at all).

To be sure, anecdotes are unreliable. A supremely creative individual such as Mozart attracts an accretion of anecdotes, not to say myths, some of which are downright false. One famous passage, quoted by Hadamard and often repeated by his readers, is probably a forgery.[5] Mozart probably *did not* write these words: '[Sometimes], thoughts crowd into my mind as easily as you could wish. Whence and how do they come? I do not know and I have nothing to do with it. Those which please me I keep in my head and hum them; at least others have told me that I do so.' Nor did he write (a few lines later), 'Then my soul is on fire with inspiration', nor (later still) 'It does not come to me successively, with various parts worked out in detail, as they will later on, but it is in its entirety that my imagination lets me hear it.'

Musicologists have rejected this spurious 'letter' since the mid-1960s. Yet, a quarter of a century later, it is still being cited without qualification by some writers on creativity (tact forbids references!). It is, of course, seductively plausible – for it fits in with the romantic and even the inspirational views, and endorses our hero-worship of Mozart to boot. (I am reminded – why? how? – of Voltaire's remark, that if God had not existed it would have been necessary to invent Him.)

The lines about conceiving the music 'in its entirety' are especially plausible. A variety of evidence suggests that Mozart, and many other H-creative people, could indeed imagine an entire conceptual structure 'all at once' (as we say). This way of putting it, like the passage from the forged letter, seems a natural way of expressing what a number of H-creative people have told us. Coleridge's notion of the poetic imagination marked this type of thinking, in which he somehow envisaged *The Ancient Mariner* as an architectural whole. And Mozart, apparently, could be simultaneously aware both of a composition's articulated inner structure and of its overall form.

But does this imply some special power, granted only to the artistic elite? Or is it a highly developed version of a power we all share?

A terrestrial explorer can survey an entire valley, seeing it simultaneously as a patchwork of roads and villages and as a glacial formation in the mountain-range. A party-goer can see, and a couturier can imagine, the structural outlines and detail of a ballgown, all at once. One can even, perhaps, imagine the song 'Where Have All the Flowers Gone?' all in a flash. – Well, perhaps. Do the flowers and the girls and the young men and the soldiers really dance together in the imagination? Or would a better description be that the successive verses and images are called up in one's mind *almost* simultaneously? When we speak of imagining the song all at once, do we merely mean that we can represent the abstract, 'circular' structure of the lyrics, perhaps with the first phrase of the melody thrown in for good measure?

Sometimes, without a doubt, we can see a hierarchical structure at several different scales-of-detail simultaneously. For instance, we can see the pattern of herringbone tweed, whose stripes are made up of smaller stripes, without having to move closer or refocus our eyes. But what about glacial valleys, or books, or folksongs – not to mention the pattern of a cantata or a symphony? Do we really experience such rich structures *all at once*?

We are facing the problem of introspection again. What may be (to you, me or even Mozart) the most natural way of describing a particular experience may not capture what actually goes on in consciousness. Still less does it identify the underlying memory-processes. Even if we do experience the valley or the folk-song all at once, the question remains as to what sorts of computation make this possible.

Our discussion of frames (in Chapter 5) is relevant here. The representation of a frame identifies both its overall structure and the items in the slots. Some slots may be unfilled (not marked as empty, but left indeterminate). Others may be filled, boringly, by pre-assigned default values. Others may have been filled, boringly or otherwise, by inspection or mental fiat. It would be impossible to represent a frame without any slots. And it would be unusual for *every* slot to be indeterminately empty (though pure mathematicians strive to define frames whose slots are as abstract as possible). Since some frames contain, or give pointers to, other frames, they can represent structures on several hierarchical levels; and again, some of the slots and sub-slots will be filled. If frames approximate some of the computational structures in our heads, then, it is not surprising that we often seem to be aware of a structure 'in its entirety'.

Similar remarks apply to other abstract schemas we have discussed, such as plans, scripts, themes or harmony. Plans involve structured computational spaces, with representations of goals, sub-goals, choice-points, obstacles and action-operators. Is it surprising, then, that your plan for getting to London tomorrow may sometimes appear 'in its entirety' in your mind? Think of the script *going to the sales*, or of the theme *escape*: don't these conjure up a number of different, yet structurally related, ideas 'all at once'? Even listening to a melody seems to involve the recognition of overall harmony 'at the same time' as accidentals, modulations or dissonance (although, as we have seen, the home-key must be established first).

These everyday examples suggest that what Mozart was able to do was of the same kind as what all of us can do – only he could do it better. We can do it for valleys, ball-gowns, trips to London and perhaps folk-songs. He could do it for symphonies.

The reason he could do it better, at least where music was concerned, is that he had a more extensive knowledge of the relevant structures. Memory, as noted earlier, stores items in the conceptual spaces within the mind. The more richly structured (and well-signposted) the spaces, the more possibility of storing items in a discriminating fashion, and of recognizing their particularities in the first place. (Broadly: the more frame-slots, the more structurally situated details.) Children, as we have seen, describe and discriminate their skills on various levels, becoming increasingly imaginative as a result. Very likely, adults do so as well.

If you could not see stripes, and mini-stripes, you could not appreciate a herringbone suit. Someone who knows nothing about glaciers cannot recognize a moraine, so cannot remember (or imagine) it either. And someone who knows nothing about tonal music cannot interpret the sounds of a Western folk-song as a melody, nor recognize a modulation or a plagal cadence. (They need not know the technical terms; but verbal labels can sometimes help to 'fix' schemas in the memory.) In short, Mozart's exceptionally well-developed musical memory was a crucial aspect of his genius.

The word 'genius' comes to mind here because Mozart was one of the very few people who have a constant, long-lasting, ability to produce H-creative ideas. Shakespeare was another, Gauss yet another. How is this possible? In other words, how can there be a degree of P-creativity so great that H-creative ideas are generated over and over again?

(We must ask the question in this way, for we saw in Chapter 3 that there can be no psychological explanation of H-creativity *as such*. What we identify as 'H-creative' depends to a large extent on historical accident and social fashion. Manuscripts are lost, and sometimes rediscovered: several unknown Mozart scores turned up recently. And even Mozart was not always revered as he is today; his music went out of fashion in Vienna after being celebrated there for years.)

Thinking can be H-creative – indeed, superlatively H-creative – in different ways. For instance, I have heard some musicians argue that Haydn was more daring than Mozart, that he challenged the musical rules more than Mozart did. If so, then Mozart's H-creativity was primarily a matter of exploring the rules to their limits (and bending and tweaking them at many unexpected points), rather than breaking them at a fundamental level. In other words, the glory of a Mozart symphony may be largely based in richly integrated musical equivalents of Dickens' exploratory use of seven adjectives to qualify one noun: we hear it with delighted amazement, for we had never realized that the relevant structural constraints had such a potential. Someone who agreed with this musical judgment might nevertheless regard Mozart as the greater genius – perhaps because his music is more diverse than Haydn's, or because it shows us the full potential of a given genre even though he did not invent it in the first place.

Whether or not an instance of H-creativity involves exceptionally radical transformation, it must involve the exploration of conceptual spaces. Accordingly, expertise is essential. If one does not know the rules (not even tacitly), one can neither break nor bend them. Or rather, one cannot do so in a systematic way.

Mere systematicity, however, is not enough. The cartoon-Einstein described in Chapter 4 was exploring a system (the alphabet), and his very next thought would have been '$e = mc^2$'. But since there is nothing about the alphabet which makes 'c' special, nothing which relates it to the speed of light or any other concept of physics, the cartoon-Einstein could not have recognized it as what he was looking for. Even everyday P-creativity requires that systematic rule-breaking and rule-bending be done in domain-relevant ways.

Consistently H-creative people have a better sense of domain-relevance than the rest of us. Their mental structures are presumably more wide-ranging, more many-levelled and more richly detailed than ours. And their exploratory strategies are probably more subtle and more powerful. Anyone can consider the negative. But they have many other (mostly domain-specific) heuristics to play with. If we could discover some of these, our educational practices might be radically improved: some of Mozart's powerful exploratory techniques, for example, might be taught to aspiring musicians.

These rare individuals, then, can search – and transform – high-level spaces much larger and more complex than those explored by other people. They are in a sense more free than us, for they can generate possibilities that we cannot imagine. Yet they respect constraints *more* than we do, not less. Where we can do nothing, or at best mentally toss a coin, they are guided by powerful domain-relevant principles on to promising pathways which we cannot even see. (Sometimes, we cannot see them until many years after they were originally traversed.)

A lifelong immersion in music lay behind Mozart's ability to abstract subtle musical structures, and to develop powerful exploratory strategies. From his very earliest years under the tutelage of his father, his life was filled with music. Pretty girls and scatological jokes aside, it seems to have been the only thing he was interested in (hence much of Salieri's exasperation).

But Mozart was not merely interested in music: he was passionate about it. In general, motivation is crucial if someone is to develop the expertise needed for H-creativity. As George Bernard Shaw put it, creativity is 'ninety per cent perspiration, ten per cent inspiration'. Even Mozart needed twelve years of concentrated practice before he could compose a major work, and much the same seems to be true of other composers.[6] In short, a person needs time, and enormous effort, to amass mental structures and to explore their potential.

It is not always easy (it was not easy for Beethoven). Even when it is, life has many other attractions. Only a strong commitment to the domain – music, maths, medicine – can prevent someone from dissipating their energies on other things. So Darwin's hypochondria, although admittedly a family trait, functioned to protect him from the tiring, time-wasting, hurly-burly of the social and scientific round. 'Resting' at home, he was not resting at all, but constantly developing and refining his ideas about evolution.

Sometimes, the emotional investment pays off in moments of pure exhilaration: on glimpsing a mathematical result (not yet a mathematical *proof*), André Ampère, as he recorded in his diary, gave 'a shout of joy'. Darwin's emotional satisfactions, one suspects, were of a less dramatic character. But satisfactions they were.

Creativity did seem to come easily to Mozart. (Poor Salieri!) And he was much more gregarious than Darwin. But even Mozart had to commit himself whole-heartedly to his chosen field. Creativity does not come cheap.

Sometimes, it comes at a very high cost indeed. Even ideas later recognized as H-creative may cause their originators more anguish than joy. Koestler tells the tragic tale of Ignaz Semmelweiss who, having discovered how to prevent puerperal fever (by washing the hands in disinfectant before attending the mothers), was exiled and eventually driven mad by the resentment of the medical profession. He remarks:

> Apart from a few lurid cases of this kind we have no record of the countless lesser tragedies, no statistics on the numbers of lives wasted in frustration and despair, of discoveries which passed unnoticed. The history of science has its Pantheon of celebrated revolutionaries – and its catacombs, where the unsuccessful rebels lie, anonymous and forgotten.[7]

These people's potentially H-creative ideas did not bring the recognition they were seeking. On the contrary, they often brought scorn, poverty and loneliness. The motivational commitment must have been exceptional for such misery to be endured.

This commitment involves not only passionate interest, but self-confidence too. A person needs a healthy self-respect to pursue novel ideas, and to make mistakes, despite criticism from others. Self-doubt there may be, but it cannot always win the day. Breaking generally accepted rules, or even stretching them, takes confidence. Continuing to do so, in the face of scepticism and scorn, takes even more.

The romantic myth of 'creative genius' rarely helps. Often, it is insidiously destructive. It can buttress the self-confidence of those individuals who believe themselves to be among the chosen few (perhaps it helped Beethoven to face his many troubles). But it undermines the self-regard of those who do not. Someone who believes that creativity is a rare or special power cannot sensibly hope that perseverance, or education, will enable them to join the creative elite. Either one is already a member, or one never will be.

Monolithic notions of creativity, talent or intelligence are discouraging in much the same way. Either one has got 'it' or one hasn't. Why bother to try, if one's efforts can lead only to a slightly less dispiriting level of mediocrity? It is no wonder if many people do not even achieve the P-creativity of which they are potentially capable.

A very different attitude is possible for someone who sees creativity as based in ordinary abilities we all share, and in practised expertise to which we can all aspire. They can reasonably hope to achieve a fair degree of P-creativity, and – who knows? – perhaps some H-creativity too. Even if their highest hopes are disappointed, they may be able to improve their imaginative powers to some significant extent.

The computational view of intelligence leaves room for such hopes. Indeed, it has led to an educational method now used in many countries: Seymour Papert's 'LOGO' programming-environment, which aims to foster skills of analysis and constructive self-criticism in children as young as five.[8] The children write simple programs telling a mechanical turtle how to draw a house, or a man, or a spiral. . . . If the turtle's house turns out not to be a proper house, the child knows that the program was somehow faulty – but the fault can be identified, and fixed. Papert claims that, as a result, children learn to analyse their own thinking as a matter of course, and gain the self-confidence both to make mistakes and to correct them; and a colleague has reported excellent results with severely handicapped children.[9] (The question of LOGO's educational effectiveness remains open, for some research suggests that the improvements do not generalize to other sorts of thinking, as LOGO-proponents assume they will.)[10]

Despite Papert's stress on self-confidence, the computational concepts he uses focus on cognitive ('intellectual') matters. The same is true of this book: computational theories of motivation were mentioned only when we discussed the understanding of stories (in Chapter 7). Until now, I have asked how novel thoughts can arise in human minds, simply taking it for granted that humans are interested in thinking them. The reason for this underplaying of motivation is that *how novel thoughts can arise* is the question that interests me most.

However, you may suspect a deeper reason. You may feel – as many people do – that motives and emotions necessarily lie outside the scope of a computational psychology. Motives, and the purposes they generate, are the origin of our actions, and are closely related to personality and the self. Emotions are seemingly opposed to rationality, since they lead us to do things without thinking – sometimes, things we would prefer to have left undone. How, then, could such aspects of the human condition be explained in computational terms?

The answer is that these phenomena will be found in *any* intelligent creature with many different, and potentially conflicting, goals. Such a creature has to be able to schedule its activities, and harmonize its many purposes, so as to optimize its success. Intelligence implies emotions, because emotions play an essential role in integrating diverse activities. And the more varied and complex a creature's goals, the more it will need higher-level structures (such as personal preferences, moral rules, and even a self-image) to organize its behaviour.[11]

For instance, goals of great urgency and importance must take precedence over the current activity, whatever that is. If we see a tiger, we run. Evolution has seen to it that we do not wait to find out whether it really is a tiger, for if our ancestors had done so we would not be here to tell the tale. Sometimes, we end up looking foolish (if the tiger was stuffed); and sometimes, our unthinking response is disastrous (if we were standing near the edge of a gorge). Occasionally, rational thought might have saved the individual concerned. But only an *automatic* interruption of the current goal-seeking activity could save the species. Quite apart from the fact that our animal-ancestors were incapable of rational thought anyway, thinking takes time – and time (in cases of urgency) is precisely what we lack. In short, the emotion of fear is not a mere feeling: it is a computational mechanism evolved for our protection.

Similarly, anxiety is a mechanism that leads us to consider more possibilities than we otherwise would, whereas confidence enables us to continue our present line of thought despite the lack of any quick success. Anxiety typically results in what computer scientists call breadth-first search, a continual hopping-about from one potential solution-path to another; confidence favours depth-first search, a calm and determined effort to follow a particular solution-path as far as necessary. Both rest on a judgment about the likelihood of success, a judgment that concerns not only the intrinsic difficulty of the problem but the person's self-image, too. Anxiety also involves estimations of the urgency and importance of the unachieved goal.

Like fear, anxiety and confidence evolved in non-human animals. They can mislead us, if the solution lies further down the mental search-tree than other animals are capable of going (so we should have stuck to our guns, instead of anxiously flitting about), or if – through overconfidence – we take undue advantage of the human capacity for sustained thinking. In either case, an unrealistic self-image can block the most suitable strategy of action. Someone who misjudges their own intellectual resources, or even their personal traits (determination, for example), may abandon a task too early or pursue it without chance of success.

Even a sudden surge of joy, on solving a difficult problem, can be seen as part of a functional mechanism: it rewards the person (or animal) for their perseverance, and releases them for other activities. But in human minds it can sometimes be dysfunctional, because people have many different values which must be satisfied – if at all – within a certain cultural milieu. Mozart, for most of his life, enjoyed the satisfactions both of creativity and of social acceptance. Semmelweiss did not; he might have been happier, overall, if he had valued the joys of creativity rather less.

These sketchy remarks tell us nothing specific about the motives of Mozart, or anyone else. Today's computational theories of motivation are painted with a broad brush. They have given us some intriguing hypotheses (and some systematic analyses of emotional terms), but they are not backed up by detailed computer models. It is difficult enough to specify how to achieve even one goal. A fully detailed account of how to deal with mutually competing purposes, in a rapidly changing and largely unfriendly world, is beyond our current understanding. But that is what motivational structures are for: the emotionless Mr Spock of *Star Trek* is an evolutionary impossibility.

Mozart, unlike Mr Spock, was fundamentally like the rest of us. But his motivational commitment was exceptional. It is hardly surprising if most people never come up with H-creative ideas. Even supposing that they have the self-confidence required, and above-average expertise, they have other fish to fry. People who live a normal life, filled with diverse activities largely prompted by other people's priorities (employers, spouses, babies, parents, friends), cannot devote themselves whole-heartedly to the creative quest. One of the ways in which Mozart was special is that he chose to do so.

'Surely,' you may say, 'expertise and commitment cannot have been all there was to it. After all, Salieri devoted his life to music too. Mozart must have been special in some other way as well.'

You may be right. Possibly, there was something about Mozart's brain which made it exceptionally efficient at picking up musical regularities, and perhaps at exploring them too. There is some evidence that musical ability (and mathematical or graphic ability too) is to some extent innate.[12]

Granted, child prodigies like Mozart are usually greatly rewarded by adults (and encouraged to practise for hours on end), so have much more opportunity to learn than other children do. Even 'ordinary' children can attain great heights, given the appropriate education. One illustration of this involves Mozart's predecessor Vivaldi, who for some years taught in a Venetian orphanage. Many of these destitute children grew up to be highly accomplished musicians, and the orphanage-concerts (at which Vivaldi's challenging new works were played, and played well) were the talk of Venice.[13]

Nevertheless, inborn factors may help certain individuals to develop the conceptual structures required. Some structures may even be inaccessible in the absence of such factors. If so, then no amount of education or commitment could suffice to form a Mozart. (Most of us, of course, would be happy enough to have a fraction of the competence of a Salieri. His music can still excite our admiration. A friend who knows Mozart's music well told me how she once switched on her wireless and was surprised, and delighted, to hear a lovely Mozart-composition which she had not heard before – it turned out to be by Salieri.)

Just what these inborn factors are, assuming they exist at all, is not known. But whatever they are, they are not supernatural. And almost certainly, they are more efficient versions of mechanisms we all share – not something profoundly different.

The capacity of short-term memory, for example, is something which might depend on fairly 'boring' facts about the brain. But the psychological implications of having a larger short-term memory might not be boring at all.

When discussing jazz-improvisation in Chapter 7, we noted that grammars of different computational power put different loads on short-term memory. Consequently, brains allowing a larger short-term memory might make certain complex structures more readily intelligible. A jazz musician might then be able to improvise new chord-sequences, as well as improvising the way in which a given chord-sequence is played. This would not explain why specifically musical structures should be favoured (though it is worth remarking that musical and mathematical ability often go together). But it might be one of several constitutional factors underlying exceptional musical prowess.

What those other factors may be is anyone's guess. Perhaps certain sorts of 'wiring' of certain groups of neurones – but which sorts, and which groups? Until we know a lot more about how the brain enables ordinary thinking and remembering to happen, we shall not be in a position to ask sensibly how Mozart's brain might have been different.

The same applies to Mozart's *mind*, to the structure-building strategies he used in composing his music. The better we understand everyday creativity, the better our chance of understanding Mozart.

In the last analysis, perhaps we never shall. Scientists will never be able to answer all possible scientific questions. And in a case like this, the scientists need the help of the musicologists. Perhaps the musicologists, no matter how hard they try, will never manage to identify all the musical structures implicit in the operas, the symphonies and the chamber music. It does not follow that Mozart's genius was *essentially* mysterious, a matter of myth rather than mechanism. Supreme puzzle, he may be. But even he was human.

11

Of Humans and Hoverflies

Holiday beaches in summertime display an awkward minuet, danced by advancing waves of spume and retreating waves of deck-chairs. As the tide rises, the deck-chairs are repeatedly moved up the shore. Only when they are safe above the high-water line do the sunbathers really relax, knowing that their territory can be encroached upon no further.

The history of science shows a similar pattern, the advance of scientific theory being matched by the retreat of anthropocentrism. Copernicus, Darwin and Freud successively challenged comfortable beliefs: that Earth is the centre of the universe, that *homo sapiens* was created in the image of God, and that people are fundamentally rational creatures. Since the Renaissance, the deck-chairs of our self-glorification have been moved several times.

Human creativity, in this scenario, lies even further up the beach than rationality does. Inspirationists and romantics lounge there at their ease, confident of being safe from science. – But is their confidence misplaced? Sometimes, after all, the high tide covers the beach, and the deck-chairs must be abandoned. Is creativity inviolable?

Well, the three intellectual revolutions cited above each showed some cherished belief to be false. Geocentrism, special creation, rational self-control: one by one, these bit the dust. If modern science were to claim that creativity is an illusion, we could sadly add a fourth example to the list.

But science claims no such thing. The previous chapters have acknowledged creativity over and over again. In brief, a scientific psychology does not deny creativity: it explains it.

To say this, however, is not enough. Many people fear that explanation *in and of itself* must devalue creativity. Forget computers, for the moment: the conviction is that *any* scientific account of creativity would lessen it irredeemably. Even an explanation in terms of brain-processes (never mind silicon-chips) would undermine our respect for creative thought.

A prime source of this common attitude is the widespread feeling that science, in general, drives out wonder. Wonder is intimately connected with creativity. All creative ideas, by definition, are valued in some way. Many make us gasp with awe and delight. We are enchanted by the water-snakes, and fascinated by the benzene-ring. To stop us marvelling at the creativity of Bach, Newton or Shakespeare would be almost as bad as denying it altogether. Many people, then, regard the scientific understanding of creativity more as a threat than a promise.

Anti-scientific prejudice of this sort is not new. William Blake had a word for it – or rather, many. 'May God us keep', he wrote, 'From Single vision & Newton's sleep!' And again:

> I turn my eyes to the Schools & Universities of Europe
> And there behold the Loom of Locke, whose Woof rages dire,
> Wash'd by the Water-Wheels of Newton: black the cloth
> In heavy wreathes folds over every Nation: cruel Works
> Of many Wheels I view, wheel without wheel, with cogs tyrannic
> Moving by compulsion each other, not as those in Eden, which
> Wheel within Wheel, in freedom revolve in harmony & peace.

To some extent, this passage is a protest against the machine-shops of the Industrial Revolution. But 'the Water-Wheels of Newton' are the wheels of science, as well as technology.

Blake was not simply objecting to machines, and the way they were changing our culture. Nor was he declaring a belief in the crystalline spheres (the wheels within wheels) of mediaeval cosmology. He was reacting against the scientistic enthusiasm that had led Alexander Pope to declare: 'God said "Let Newton be!", and all was light.' For Blake, Newton's light made only *single* vision possible. Matters not dealt with by natural science, such as freedom and harmony, were insidiously downgraded and ignored – even tacitly denied.

Science withstood Blake's attacks, and grew apace. It spawned many new theories and many, many, new facts. But reservations about the scientific world-view remained, and remain to this day.

Some doubters wielded the weapon of humour: Dickens, in the 1830s, mocked the infant British Association for the Advancement of Science (widely referred to as 'the British Ass') in his *Mudfog Papers*. The anti-scientific writers of the 1960s 'counter-culture' – Theodore

Roszak, for one – were more passionate, if less witty. Like Blake, they criticized science for what they saw as its mechanistic denial of freedom (a criticism discussed later). And they called, specifically, for a return to religious reverence, or wonder, if not to theological dogma.

A nineteenth-century contributor to the *Athenaeum* put it in a nutshell. He had shared a stage-coach with some scientific worthies travelling to the 1834 meeting of the British Association in Edinburgh. This is how he described the experience:

> We entered Scotland over the Cheviot Hills. Their appearance attracted the notice of all, and it was soon evident that our fellow-travellers were members of the Association, full of their respective subjects, eager to impart and receive information. . . . *Science destroyed romance* – the field of Chevy Chase scarce elicited a remark – the cross marking the spot where Percy fell was observed by one of the geologists to belong to the secondary formation; the mathematician observed that it had swerved from the perpendicular, and the statisticians began a debate on the comparative carnage of ancient and modern warfare. [Italics added.]

We have all encountered scientifically minded individuals with manners bad as these, full of their own knowledge but ignorant of history and blind to the beauties of landscape. (We have all met personally obnoxious artists, too.) But does science *necessarily* destroy romance?

Sometimes, to be sure, it does. Science is fundamentally opposed to the romance of superstition (which includes the inspirational and romantic 'theories' of creativity). When our wonder is based on ignorance, error or illusion, it must fade in the light of understanding.

But science can lead in turn to a new form of wonder, which is not so easily destroyed.

A friend who is an engineer once told me how, as a very young child, he was for a while utterly fascinated by circles. He would collect circular things – coins, bottle-tops, tins – which he kept in his toy-cupboard, and which he used to draw circles of many different sizes. One day his parents told him that there is an instrument that can draw any circle whatever (up to a certain size). He wondered greatly at this idea, and could hardly wait to receive this marvel as a gift. He thought of it as some sort of magically changing item that could transform itself into equivalents of all the different objects he had collected in his cupboard.

Then he was given a compass. He was horribly disappointed, for there was nothing magical at all about the compass. It was boringly simple, and its 'power' was transparent even to an infant. He still remembers this day of disillusion as a traumatic event in his childhood.

Today, however, he has the maturity to see that the compass was indeed wonderful – and the mathematical principle it embodied, even more so. Its simplicity (which can generate many superficially varying cases) is 'boring' only to those who feel that Baroque confusion is a necessary mark of the wonderful. Even Blake did not believe this: hence his reference to harmony.

Mysteries were contrasted, in Chapter 1, with puzzles. Physics, chemistry and molecular biology have already transmuted many mysteries into puzzles, and solved them to boot. Now, psychology is helping us to understand how the previously mysterious behaviour of humans and other animals is possible.

In some cases, this added understanding makes us react much as the infant engineer did to the compass. That is, the newly discovered simplicity drives out our wonder, stifles our sense of awesome mystery, leaving us only with the brute facts of science.

Consider the hoverfly, for example. A hoverfly is able to meet another hoverfly in mid-air – which is just as well, since they need to be at the same place if they are to mate. How does this mid-air meeting come about?

One might assume that the hoverfly does something similar to what a person does on recognizing a friend across a city-square: altering direction immediately, and adjusting their path as necessary if the friend suddenly swerves. Sentimentalists would expound on the wonders of nature, as illustrated by the marvellous powers of the humble hoverfly. More sober souls (given the assumption in question) might feel some sympathy for such a view. It turns out, however, that this assumption about how the hoverfly manages its social life is false.

On closer examination, there is nothing like the flexible selection and variation of pathways that are involved in truly intelligent friend-seeking action. For the fly's flight-path is determined by a very simple and inflexible rule. This rule, which is hardwired into the insect's brain, transforms a specific visual signal into a specific muscular response. The fly's change of direction depends on the particular approach-angle subtended by the target-fly at the time. The creature, in effect, always assumes that the size and velocity of the seen target (which may or may not be a fly) are those corresponding to hoverflies. When initiating its new flight-path, the fly's angle of turn is selected on this rigid, and fallible, basis. Moreover, the fly's path cannot be adjusted in midflight, there being no way in which it can be influenced by feedback from the movement of the target animal.

This evidence must dampen the enthusiasm of anyone who had marvelled at the similarity between the hoverfly's behaviour and the ability of human beings to intercept their friends. The hoverfly's intelligence has been demystified with a vengeance, and it no longer seems worthy of much respect.

To be sure, one may see beauty (like the beauty of the compass) in the evolutionary principles that enabled this simple computational mechanism to develop, or in the biochemistry that makes it function. But the fly itself cannot properly be described in anthropomorphic terms. Even if we wonder at evolution, and at insect-neurophysiology, we can no longer wonder at the subtle mind of the hoverfly.

Many people fear that this disillusioned denial of intelligence in the hoverfly is a foretaste of what scienice will say about our minds, too. But this is a mistake. The mind of the hoverfly is much *less* marvellous than we had imagined, so our previous respect for the insect's intellectual prowess is shown up as mere ignorant sentimentality. But computational studies of thinking can increase our respect for human minds, by showing them to be much *more* complex and subtle than we had previously recognized.

Think of the many different ways (sketched in Chapters 4 and 5) in which Kekulé could have seen snakes as suggesting ring-molecules. Consider the rich analogy-mapping in Coleridge's mind, which drew on naval memoirs, travellers' tales, and scientific reports to generate the shining water-snakes that swam through Chapter 6. Or remember the mental complexities (outlined in Chapter 7) underlying a plausible, and grammatically elegant, story. Even relatively simple computational principles, such as the jazz-program's rules for melodic contours, can – like the compass – have surprisingly rich results. A scientific psychology, by identifying the mental processes involved in examples like these, can help us to appreciate just how wonderful the human mind is.

Admittedly, poets and novelists have long had an intuitive sense of some of the psychological subtleties concerned. Consider Proust's insightful depiction of memory, or Coleridge's (and Livingston Lowes') comments on mental association. Theoretical psychologists such as Freud relied on similar insights, when discussing symbolism in the dreams and dramas of everyday life. But such notions have remained literary and intuitive, rather than scientifically rigorous. Moreover, even Freud underestimated the degree of complexity of the mental processes he described. So did Koestler, in his attempts to define 'the bisociation of matrices'.

To cease to wonder at creativity because it had been explained by science would be to commit 'the fallacy of the compass'. The temporary disaffection of the infant engineer was irrational and unnecessary. Now, he still values circles for the beauty of their superficial form. But he also appreciates the underlying mathematical principle, which enables them to be generated by anyone with a compass to hand. He has gained, not lost.

A scientific psychology, then, allows us plenty of room to wonder at Mozart, and even at Grandpa's jokes. Much as geology leaves the Cheviot Hills as impressive as ever, and Chevy Chase as poignant, so psychology leaves poetry in place. Indeed, it adds a new dimension to our awe on encountering the water-snakes, or the theory of the benzene-ring.

Darwin made a similar point about biology. Rather than denying our wonder at God's creation, he said, evolutionary theory can increase it:

> [I think it] an idea from cramped imagination, that God created the Rhinoceros of Java and Sumatra, that since the time of the Silurian he has made a long succession of vile molluscous animals. How beneath the dignity of him, who is supposed to have said 'let there be light,' and there was light.

'The more magnificent view', he said, is that all these creatures, along with their more aesthetically appealing cousins, have been produced by 'the body's laws of harmony'.

'Ah,' someone might say, on reading Darwin's words above, 'but there's the rub! The *body's* laws of harmony are one thing. Brains may well have creative powers. But computers are quite another matter!'

This objection can be interpreted in at least three ways. These recall the first, second and last Lovelace-questions, respectively.

It may mean that computers are utterly irrelevant to human creativity, and cannot possibly help us to understand it. – It may mean that computer-performance could never match ours, that there will never be a computerized Chopin or Donne. – Or it may mean that computers, unlike people, cannot *really* be creative.

Let us consider each of these, in turn.

As regards the first interpretation (that computers are utterly irrelevant to human creativity), it is *computational concepts and theories*, not computers as such, which are crucial for psychology. Computational psychology tries to specify the conceptual structures and processes in people's minds. The material embodiment of a particular computation need not be silicon, or gallium arsenide, or anything else dreamt up by computer-engineers. It may be good old-fashioned neuroprotein, working inside human heads.

Computers are very useful nevertheless, because their programs are effective procedures. If a jazz-program produces acceptable music, we *know* that essentially similar computations grounded in the brain could be the source of real, live jazz.

Anyone who doubts whether the brain does things in quite that way may be right. But they should provide specific evidence, not vague intuitive prejudice, to support their case. (Johnson-Laird does this when he argues that limits on short-term memory make it impossible for jazz-melodies to be improvised by hierarchical grammars, and that limits on long-term memory prevent them being strings of motifs.)

Preferably, the sceptics should offer an alternative psychological theory, with equal clarity and more appropriate generative power. If they manage to do so, then the original hypothesis will have been scientifically fruitful, even though rejected (a common fate of creative ideas in science). If they do not, then the contested theory stands, as the most promising explanation so far.

Promising explanations abound, as we have seen throughout this book. Even motivation and emotion have been analysed in computational terms, though these theories are relatively sketchy as yet. Despite such theoretical gaps, there are many ways in which a computational approach can help to explain human creativity. For conceptual spaces – and the ways in which they can be mapped, explored and transformed – are made more precisely intelligible by thinking of them in computational terms.

The concept of a generative system, for instance, enables us to understand how ideas can appear which, in an important sense, *could not* have appeared before. It helps us to focus on styles of thinking in science and the arts, and to analyse how they can be changed in more or less radical fashions. These questions are the concern of musicologists, literary critics and historians of art and science, whose insights we need if we are to understand creativity. Many of these insights are highly subtle, and not easily expressed in precise terms. But our discussions of harmony, jazz and line-drawing – not to mention scientific creativity of various sorts – showed that computational psychology can already say something specific about such matters.

Computer-modelling enables us to test our psychological theories by expressing them as effective procedures, as opposed to verbal theories, vaguely understood. It helps us to see how complex, yet how constrained, is the appreciation of a piece of music - or its improvisation. It replicates some scientific discoveries from centuries past, and sharpens our sense of what certain methods of reasoning *can* and *cannot* achieve. And it confirms the insight that changing the representation of a problem may make it much less difficult, for identifiable reasons.

The notion of heuristic search, controlled in specifiable ways, helps us to understand how a wide range of H-creative ideas could have arisen. Scripts, and related concepts, indicate some ways in which knowledge may be organized within the mind, and help us to appreciate the many varied constraints involved in writing a story. The 'unnaturalness' of step-by-step programming is largely irrelevant: even though the brain often uses heuristics, scripts and frames in parallel, it may be exploring conceptual spaces whose dimensions are those embodied in sequential models.

The richness and subtlety of 'ordinary' psychological abilities are highlighted by this approach. We can see, much more clearly than Coleridge or Koestler could, how everyday functions of memory and comparison might underlie creative thinking. And we can see how even the mundane task of describing a game of noughts-and-crosses requires the integration of many different constraints, not only the rules of grammar.

Moreover, computational psychology is to some extent inspired by ideas about the brain. Connectionist models give us some grip on the fleeting psychological processes involved in poetic imagery, scientific analogy and serendipity. They define mechanisms capable of global recognition, where no single feature is necessary but many features are sufficient. And they show how a mind (and even a computer) can accept imperfect pattern-matches, and retrieve an entire associative complex on being given a fragment of it. Snakes and water-snakes, given their creative contexts of molecules and mariners, are made less mysterious accordingly.

Connectionism must be combined with other kinds of computational theory, in order to model the deliberate thinking found in the evaluation phase (and often in the preparatory phase, too). One of the most active research-areas at present is the design of 'hybrid' systems, combining the flexible pattern-matching of connectionism with sequential processing and hierarchical structure. A psychological theory of creativity must explain both types of thinking, and how they can coexist in a single mind.

Many other examples have been given in support of my claim that the first Lovelace-question merits the answer 'Yes'. If you are not convinced by now, nothing I could say at this point would help. With respect to the first interpretation of 'Computers are another matter!', then, I rest my case.

Under the second interpretation (that computer performance could never match ours), the objection is less easily countered.

If future programs are to match all our creative powers, then future psychology must achieve a *complete* understanding of them. Our knowledge of the human mind, and of its potential, has been greatly increased by the computational approach. Neuroscience will doubtless improve our understanding of these matters, too. But there are many still-unanswered questions about human thinking, most of which have not even been asked. Why should anyone believe that we shall find answers to every single one of them?

The conceptual space of theoretical psychology is too large for us to explore every nook and cranny. Perhaps those never-to-be-explored regions hold some secrets of Chopin's music and Donne's poetry. Likewise, some physical diseases may never be understood: it does not follow that physiology and molecular biology are a waste of time. Science cannot answer, or even ask, all possible scientific questions.

Even if it could, scientists would not want to waste their time, and money, integrating all these explanatory principles within a single computer-model. (Combining BACON with its four P-creative cousins is child's play by comparison.) Some special-purpose computer systems will very likely be built, to do for other sciences what DENDRAL did for chemistry – indeed, much more. But there are easier, and more enjoyable, ways of generating new wits and new poets. As Koestler said,

> The difficulty of analysing the aesthetic experience is not due to its irreducible quality, but to the wealth, the unconscious and non-verbal character of the matrices which interlace in it, along ascending gradients in various dimensions.[1]

The preceding chapters have suggested how many, and how very various, these dimensions are. To put all of them into one AI-model would in practice be impossible.

Moreover, we have seen that human creativity often involves highly idiosyncratic experiences (Proust's madeleine, for one). But theoretical psychology is concerned with general principles, not personal biographies or gossip. Even when a psychologist does discuss detailed personal evidence (such as Darwin's notebooks, within which one can see the concept of evolution being gradually developed), these are used as grist for the theoretical mill.

A computer-model embodying general psychological principles has to be given some content, something idiosyncratic to chew on. It may be tested on Socrates' philosopher–midwife analogy, as we have seen. But it would not be worth anyone's while, even assuming it were feasible, to build in all of Socrates' knowledge and experience – including his marital tiffs with Xanthippe and his filial bonds with the real midwife, Phainarete. Unless that were done, however, the rich texture of Socrates' thought would not have been modelled in detail, and his creativity could not be fully captured. This has nothing to do with Socrates' greatness: your creativity, and your next-door-neighbour's, cannot be exhaustively modelled either.

Future computers will perform creatively, up to a point. (Arguably, some already do.) So the answer to the second Lovelace-question is a guarded 'Yes'. But to expect a computer-model to match the performance of Chopin or Donne is unrealistic. Even to mimic the wit and wisdom of a schoolgirl's letter to her best friend is probably too great a challenge. To await a computerized Shakespeare is to wait for Godot.

The third interpretation of 'Computers are quite another matter' raises very different – and very difficult – issues. It holds computers to be intrinsically incapable of genuine creativity, *no matter how impressive their performance may be.*

For the purpose of argument, let us assume that computers could one day appear to be as creative as we are. They might have their blind spots, as we do too: sneezing and chilblains (two examples drawn from Chapter 1) might be understood by them only in a theoretical, book-learnt, way. But they would produce countless ideas – cantatas, theorems, paintings, theories, sonnets – no less exciting than ours. And they would do so by means of computational processes like those which, according to theoretical psychologists, go on in human heads.

Even so, this objection insists, they would not really be creative. Indeed, they would not really be intelligent. Artificial intelligence (on this view) is analogous not to artificial light, but to artificial five-pound notes. Far from being an example of the same sort of thing, it is something utterly different – and to pretend otherwise is fraudulence.

'Well, the question's clear enough!' you may say. 'What's your answer to it?' Not so fast! We are dealing, here, with the fourth Lovelace-question: could a computer *really* be creative? And this question is not clear at all. Indeed, the answer 'No' may be defended in (at least) four different ways. Let us call these the brain-stuff argument, the empty-program argument, the consciousness argument and the non-human argument.

The brain-stuff argument relies on a factual hypothesis: that whereas neuroprotein is a kind of stuff which can support intelligence, metal and silicon are not. The empty-program argument makes a philosophical claim: that all the symbols dealt with by a computer program are utterly meaningless to the computer itself. The consciousness argument claims, similarly, that no computer could conceivably be conscious. And the non-human argument insists that to regard computers as *truly* intelligent is not a mere factual mistake (like saying that a hoverfly's blood is exactly the same as ours), but a moral absurdity.

We must consider each of these four arguments, for someone who insists that computers cannot *really* be creative may have any one (or even all) of them in mind.

Hoverflies, as we have seen, are not very bright. According to the brain-stuff argument, however, even hoverflies have a greater claim to intelligence than computers do. For the measure of computer-intelligence is precisely zero.

More accurately, the claim is that computers made of inorganic materials must be for ever non-intelligent. Only 'biological' computers, built from synthetic or naturally-occurring organic compounds, could ever achieve real thinking. Hoverflies, who share the same genetic code as we do, and whose bodies contain chemicals broadly comparable to ours, can perhaps claim a fleeting antenna-hold on intelligence. But computers cannot. In a word: no biochemistry, no creativity.

The main factual assumption driving this argument may, conceivably, be true. Possibly, computers are made of a sort of material stuff which is incapable of supporting intelligence. Indeed, neuroprotein may be the *only* substance in the universe which has this power.

Then again, it may not. Even carbon-strings and benzene-rings may not be necessary: there may be creative intelligences on Mars or Alpha Centauri, with alien chemicals filling their heads. Science does not tell us this is impossible.

'Never mind Martians!' someone may say. 'We're talking about computers. It's obvious that metal and silicon can't support intelligence, whereas neuroprotein can.'

But this is not obvious at all. Certainly, neuroprotein does support intelligence, meaning and creativity. But we understand almost nothing of how it does so, *qua* neuroprotein – as opposed to some other chemical stuff. Indeed, insofar as we do understand this, we focus on the neurochemistry of certain basic *computational functions* embodied in neurones: message-passing, facilitation, inhibition and the like.

Neurophysiologists have discovered the 'sodium-pump', for instance. This is the electrochemical process, occurring at the cell membrane, which enables an electrical signal to pass (without losing strength) from one end of a neurone to the other. And they have studied the biochemistry of neurotransmitters, substances (such as acetylcholine) that can make it easier – or harder – for one nerve-cell to cause another one to fire. In a few cases, they have even been able to say something about how a cell's chemical properties (and connectivities) enable it to code one sort of information rather than another: picking up colours or light-intensity gradients, for instance, or sounds of varying pitch.

If we could not recognize sounds of varying pitch, we could not appreciate music. Heuristics of harmony, like those described earlier, can be applied only if one can hear (for instance) that one note is a semitone lower than another. If the neurophysiologist can tell us not only which auditory cells enable us to do this, but what chemical processes are involved, all well and good. But the neurochemistry is interesting only to the extent that it shows how it is possible (in human heads) to compute tonal relationships. *Any other chemistry would do*, provided that it also enabled harmonic intervals to be computed.

Likewise, we need to see lines if we are to draw (or appreciate drawings of) acrobats. But any chemistry would do in the visual system, so long as light-intensity gradients could be identified by means of it. Again, any chemical processes would do at the cell-membrane, and at the synapse, provided that they allowed a nerve-cell to propagate a message from one end to the other and pass it on to neighbouring neurones.

Recognizing sounds and lines are computational abilities which some computers already possess. Our own mental life, however, contains much more than tonal harmony and line-drawing. Conceivably, there may be other sorts of computation, going on inside human heads, which simply cannot be embodied in anything made of metal and silicon. But we have no specific reason, at present, to think so. Conceivably, too, only neuroprotein can implement the enormous number of stable yet adaptable structures involved in human thought, and/or of doing so in manageable space and time. Again, however, we have no particular reason to think so (and what is to count as 'manageable'?).

The fact that we cannot see how metal and silicon could possibly support 'real' intelligence is irrelevant. For, intuitively speaking, we cannot see how neuroprotein – that grey mushy stuff inside our skulls – can do so either. *No* mind–matter dependencies are intuitively plausible. Nobody who was puzzled about intelligence (as opposed to electrical activity in neurones) ever exclaimed '*Sodium* – of course!' Sodium-pumps are no less 'obviously' absurd than silicon-chips, electrical polarities no less 'obviously' irrelevant than clanking metal. Even though the mind–matter roles of sodium-pumps and electrical polarities are scientifically compelling, they are not intuitively intelligible. On the contrary, they are highly counter-intuitive.

Our intuitions will doubtless change, as science advances. Future generations may come to see neuroprotein – and perhaps silicon, too – as 'obviously' capable of embodying mind, much as we now see biochemical substances in general as obviously capable of producing other such substances (a fact regarded as intuitively absurd, even by most chemists, before the synthesis of urea in the nineteenth century). As yet, however, our intuitions have nothing useful to say about the material basis of intelligence.

In sum, the brain-stuff argument is inconclusive. It reminds us that computers made of non-biological materials may be incapable of real creativity. But it gives us no reason whatever to believe that this is actually so.

You may have come across the empty-program argument on your TV-screen or radio, or in your newspaper. For a recent version of it, based on John Searle's intriguing fable of the Chinese Room, has figured in the international media.[2]

Searle imagines himself locked in a room, in which there are various slips of paper with doodles on them. There is a window through which people can pass further doodle-papers to him, and through which he can pass papers out. And there is a book of rules (in English) telling him how to pair the doodles, which are always identified by their shape. One rule, for example, instructs him that when *squiggle-squiggle* is passed in to him, he should give out *squoggle-squoggle*. The rule-book also provides for more complex sequences of doodle-pairing, where only the first and last steps mention the transfer of paper into or out of the room. Searle spends his time, while inside the room, manipulating the doodles according to the rules.

So far as Searle-in-the-room is concerned, the *squiggles* and *squoggles* are meaningless. In fact, however, they are characters in Chinese writing. The people outside the room, being Chinese, interpret them as such. Moreover, the patterns passed in and out at the window are understood by them as questions and answers respectively: the rules ensure that most of the questions are paired, either directly or indirectly, with what they recognize as a sensible answer. Some of the questions, for example, may concern the egg foo-yong served in a local restaurant. But Searle himself (inside the room) knows nothing of that. He understands not one word of Chinese. Moreover, he could never learn it like this. No matter how long he stays inside the room, shuffling doodles according to the rules, he will not understand Chinese when he is let out.

The point, says Searle, is that Searle-in-the-room is acting as if he were a computer program. He is all syntax and no semantics, for he is performing purely formal manipulations of uninterpreted patterns. It is the shape of *squiggle-squiggle*, not its meaning, which makes him pass a particular piece of paper out of the window. In this sense, his paper-passing is like the performance of a 'question-answering' program, such as the restaurant program or BORIS (mentioned in Chapters 5 and 7 respectively). But Searle-in-the-room is not *really* answering: how could he, since he cannot understand the questions?

It follows, according to Searle, that a computational psychology cannot explain how it is possible for human beings to understand meanings. At best, it can explain what we do with meanings, once we understand them. (So Searle's answer to the *first* Lovelace-question – whether computational concepts can help us understand human creativity – would be 'Perhaps, but not at a fundamental level.') *A fortiori*, no program could ever give a computer the ability to understand. Complexity does not help, since all it does is add more internal doodle-matchings: no mega-BORIS of the future will really understand stories. The answer to the last Lovelace-question, on this view, must be 'No'.

If you are not immediately convinced by Searle's argument, your first response may be to object that he has cheated. Of course Searle-in-the-room will never learn Chinese, for he is not causally plugged in to the world. No one can really understand what egg foo-yong is without being able to see it, smell it and poke at it with chopsticks. A computer capable of constructing real meanings would need to be more than a VDU-screen attached to a teletype – which is the sort of computer that Searle-in-the-room is mimicking.

Thus far, Searle would agree with you. But he responds by imagining that the room, with him (and a new rule-book) in it, is placed inside the skull of a giant robot. Now the *squiggle-squiggles* are caused by processes in the robot's camera-eyes, and the *squoggle-squoggles* cause levers in the robot's limbs to move. Accordingly, the egg foo-yong can be picked up and transferred into the robot's mouth. But, Searle argues, Searle-in-the-robot has no idea what it is to move a chopstick, and does not know an egg foo-yong when the camera records it.

Notice that Searle, in these imaginary situations, must understand the language (English, we are told) of the rule-books. Notice, too, that the rule-books must be at least as detailed as AI-programs are. They must include rules for using language grammatically (no simple matter, as we have seen), and rules covering vision and motor control. In short, in order to write the rule-books one would need a powerful computational psychology, involving theoretical concepts like those discussed in previous chapters. Whether English is sufficiently precise for this task is – to put it mildly – doubtful. What would be needed is something like an AI-programming language. And this language would have to be taught to Searle before his incarceration, for the story depends on his being able to understand it. We shall return to this point later.

Searle's crucial assumption is that computer programs are semantically empty. That is, they consist of abstract rules for comparing and transforming symbols not in virtue of their meaning, but merely by reference to their shape, or form. The so-called symbols are not really symbols, so far as the computer is concerned. To the computer, they are utterly meaningless – as Chinese writing is to Searle-in-the-room. Human beings may interpret them as concepts of various kinds, but that is another matter.

That programs are empty in this sense is taken for granted by Searle. But it is true only if we think of them in a particular way.

Remember the necklace-game: although I described it in terms of red, white and blue beads, I said that it was based on Hofstadter's 'pq-system'. The pq-system is a logical calculus, whose rules generate strings of letters. As such, it has syntax (only letter-strings with certain structures are allowed), but no semantics. Considered purely as a logical calculus, wherein the letters are not interpreted in any way, it is meaningless. But considered as a set of rules for necklace-building (or for doing addition), it is not.

Remember, too, the distinction (made in Chapter 3) between the two sorts of computational *coulds*. One is a timeless, abstract sense: *could* this set of numbers (successive squares, for instance) be generated, in principle, by such-and-such a mathematical rule? The other is a procedural, and potentially practical, sense: *could* this rule actually be used to produce the relevant set of numbers? The former question asks whether a certain structure lies within a particular conceptual space. The latter asks whether the space can be explored, in a specific manner, so as to find it. Both types of question can be applied to computer programs.

One can, for some purposes, think of a computer program as an uninterpreted logical calculus, or abstract mathematical system. This is useful if one wants to know whether it is capable, in principle, of producing certain abstractly specified results. (For example: could it get stuck in an infinite loop, or could it distinguish grammatical sentences from nonsensical word-strings?) But one must not forget that a computer program is *a program for a computer*. When a program is run on suitable hardware, the machine *does* something as a result. It builds computational necklaces, as one might say.

There is no magic about this, just as there is no magic about the hoverfly's finding its mate. Input-peripherals (teletypes, cameras, sound-analysers) feed into the internal computations, which lead eventually to changes in the output-peripherals (VDU-screens, line-plotters, music-synthesizers). In between, the program causes a host of things to happen, many symbols to be manipulated, inside the system itself. At the level of the machine code, the effect of the program on the computer is direct, because the machine is engineered so that a given instruction elicits a unique operation. (Instructions in high-level languages must be converted into machine-code instructions before they can be obeyed.)

A programmed instruction, then, is not merely a formal rule. Its essential function (given the relevant hardware) is to make something happen. Computer programs are not 'all syntax and no semantics'. On the contrary, their inherent causal powers give them a toehold in semantics.

In much the same way, the causal powers of the hoverfly endow its 'mind' with primitive meanings. But because the fly's internal computations are not complex enough to enable it to plan, or even to react to changes in another hoverfly's flight-path, its meanings are neither diverse nor highly structured. It is, as we noted before, not very bright. Indeed, its computational powers are so limited that we may refer to its mind only in scare-quotes, as I just did.

Because Searle assumes the utter emptiness of programs, he draws the wrong analogy. At base, a functioning program is comparable with Searle-in-the-room's understanding of English, not Chinese.

A word in a language one understands is a mini-program, which causes certain processes to be run in one's mind. Searle sees Chinese words as meaningless doodles, which cause nothing (beyond appreciation of their shape) to happen in his mind. Likewise, he hears Chinese words as mere noise. But he reacts in a very different way to words from his native language. English words trigger a host of computational procedures in his head: procedures for parsing grammatical structures, for accessing related ideas in memory, for mapping analogies, for using schemas to fill conceptual gaps . . . and so on. And some English words set up computations that cause bodily actions (for example, 'Pass the slip of paper out of the window').

To learn English is to set up the relevant causal connections, not only between words and the world ('cat' and the thing on the mat) but between words and the many non-introspectible processes involved in interpreting them. The same applies, of course, to Chinese (which is why Searle-in-the-room's rule-book must contain rules for parsing, interpreting and constructing sentences in Chinese). Moreover, the same applies to any new (AI-like) language which the people outside the room may teach Searle, in order that he should understand the rule-book. He would have to learn to react to it (automatically) as the electronic machine is engineered to do.

Where does our discussion of the empty-program argument leave us, with respect to the fourth Lovelace-question? Could a computer *really* understand anything at all?

Certainly, no current 'question-answering' program can really understand any natural-language word. Too many of the relevant causal connections are missing. BORIS, for example, does not really understand why (in the story mentioned in Chapter 7) Paul phones his friend Robert for legal advice on discovering his wife's infidelity. It does not really know what a telephone is, still less what lawyers, friendship and jealousy are.

But BORIS does have the beginnings of an understanding of what it is to compare two symbols, and of what it is to plan and to parse. You may want to say that it cannot really do these things, either – that it cannot really interpret plans, nor really parse sentences. But the complaint that it cannot parse is surely dubious. As for plans, BORIS has access to some of their abstract structural features, such as means–end relationships and the possibility of cooperation or sabotage at certain points. In short, the understanding that BORIS possesses is of a very minimal kind: how to compare two formal structures, for example, or how to build a new one by using certain hierarchical rules.

Perhaps you feel that the 'understanding' involved in such a case is so minimal that this word should not be used at all? So be it. For the purpose of explaining *human* creativity, we need not ask, 'When does a computer (or, for that matter, a hoverfly) understand something?' Moreover, this question is ill-advised even if we are interested in computers as such, for it misleadingly implies that there is some clear cut-off point at which understanding ceases. In fact, there is not.

The important question is not 'Which machines can understand, and which cannot?', but 'What things does a machine – whether biological or not – need to be able to do in order to be able to understand?'[3] These things are very many, and very various. They include not only responding to and acting on the environment, but also constructing internal structures of many different kinds.

Hoverflies can do relatively few of these things, which is why their intellectual powers are unimpressive. Current computers can do a few (others) too, but they lack the ways of situating themselves in a meaningful external world which even hoverflies possess. Robots with efficient senses and motor organs would be more like hoverflies. But they would need many different world-related goals, and ways of constructing new ones, to be likened to mammals.

In general, the more types of conceptual space that can be built, and the more flexibly and fruitfully they can be explored and transformed, the greater the understanding – and the greater the creativity. Computational psychology has provided a host of theoretical ideas with which to map the conceptual spaces constructed within human minds. Moreover, *pace* Searle, it even helps us to see how creative understanding is possible at all.

The third way in which people commonly deny the possibility of 'real' creativity in computers is to appeal to the consciousness argument. 'Creativity requires consciousness,' they say, 'and no computer could ever be conscious.'

We have seen, time and time again, that much – even most – of the mental processing going on when people generate novel ideas is not conscious, but unconscious. The reports given by artists, scientists and mathematicians show this clearly enough. To that extent, then, this argument is misdirected.

But it does have some purchase, for the concept of creativity includes the notion of positive evaluation, which involves the deliberate examination and modification of ideas. (Remember Kekulé's adjustments to the benzene-ring, in the light of valency?) In the four-phase account of Poincaré and Hadamard, this is the phase of 'verification'. Our discussion (in Chapter 4) of the development of self-conscious reflection in children indicated how crucial to creativity this is. To be creative, one must be able to map, explore and transform *one's own mind*.

It follows that computers can be creative only if, in this sense, they can be conscious. 'Well then,' you may say, 'our imaginary objector must be right. After all, no mere tin-can computer could be *conscious*!'

Wait a minute. 'Conscious' (like 'consciousness') is a word with several different meanings. The sort of consciousness that is essential for creativity, because it is involved in the very definition of the term, is self-reflective evaluation. A creative system must be able to ask, and answer, questions about its own ideas. Are elliptical orbits – or benzene-rings, or modulations to a distant key – acceptable, given the relevant constraints? Are they more illuminating than the previous ways of thinking, more satisfactory than any alternatives currently in mind? Is it interesting – or even intelligible – to refer to thrushes' nests as 'little blue heavens', or to compare sleep with a knitter, or a bath?

Such questions can certainly be asked by computer programs. Lenat's program AM, for example, asks whether its newly generated categories (multiplication, primes, maximally divisible numbers . . .) are mathematically 'interesting'. If it decides that they are, it explores them further (as well as reporting them in its printout). DALTON and related models of scientific discovery examine their newly produced conceptual structures in various ways, representing verification of different kinds. And the combined ARCS–ACME system can reflectively assess the strength of the literary or scientific analogies it has generated. Presumably, the computer systems of the future will be able to represent, and so to examine, their novel ideas in even more subtle ways. In *this* sense of the term, then, there seems no reason in principle why a computer could not be conscious.

There are several other senses of 'conscious'. Some of them can (like self-reflective thinking) be accepted within a computational psychology. And some puzzling facts about consciousness can even be explained in computational terms.

For example, consider the strange non-reciprocal co-consciousness sometimes found in clinical cases of 'split personality'. In one famous case, the sexually aware personality known as Eve Black had conscious access to the thoughts of the demure Eve White, and even commented spitefully on them to the psychiatrist. It was as though Eve Black could introspect Eve White's mind – except that Eve Black vehemently denied that it was *her own* mind she was looking into, so 'introspection' hardly seems the right word. But Eve White, like the rest of us, had conscious access to nobody's thoughts but her own. Consequently, Eve Black could – and did – play highly embarrassing tricks on Eve White, gleefully recounting them later to the doctor.[4]

How are such things possible? Taking the notion of conscious access for granted (as theoretical psychologists usually do, if they use the notion at all), this sort of psychological phenomenon can be understood in computational terms. Broadly, we can think of the two 'personalities' as different modules of one overall computational system, alternately controlling the same motor facilities and sensory apparatus (the patient's body), and having different degrees of access to and control over each other's current processing or memory-store. This does not explain why the psychological dissociation happened in the first place, what events in the person's life-history caused it. But the problem here was to understand *how it is possible at all*, not what triggered it.

In considering our original question, however, the notion of conscious access cannot be taken for granted. It involves, among other things, the problematic idea of felt experience, or sensation: that introspective *something* that is so remarkably difficult to describe, but which we all know in our own case. (Just try pinching your arm!) What of that?

Well, what of it? It's not obvious that consciousness in *that* sense is essential to creativity. It's not obvious, either, that the various *somethings* can actually be distinguished independently of their causal relationships with other internal processes and with things in the world outside. If they cannot, then perhaps some conceivable computer could (as we say) experience them. (A machine capable of feeling would doubtless be very unlike the computers we know today, but that is another matter.)

Our previous discussion of the brain-stuff argument remarked that the relation of mental phenomena to neuroprotein, as such, is utterly mysterious. So why shouldn't some future tin-can have feelings and sensations, too? Why shouldn't there be *something which it is like to be* that computer, just as there is *something which it is like to be* you or me – or a bat?[5] Admittedly, it seems intuitively unlikely, perhaps even absurd. But intuitions are not always reliable.

The crucial point here is that we understand so little about this particular sense of consciousness that we hardly know how to speak about it, still less how to explain it. When we say with such confidence that *we* have consciousness, we do not know what it is that we are saying. In these circumstances, we are in no position to prove that no computer could conceivably be conscious.

In sum, *if* creativity necessarily involves conscious experience over and above the self-reflective evaluation of ideas, and *if* no computer could have conscious experience, then no computer could 'really' be creative. But these are very iffy ifs. The question must remain open – not just because we do not know the answer, but because we do not clearly understand how to ask the question.

What of the fourth way of denying 'real' creativity to computers: the non-human argument? Unlike the brain-stuff argument, this is not a scientific hypothesis. And unlike the empty-program argument and the consciousness argument, it is not a disinterested philosophical debate. Rather, it is the adoption of a certain attitude towards computers, an uncompromising refusal to allow them any social roles like those enjoyed by people.

I suggested in Chapter 7 that if a future version of AARON were to draw acrobats with 'triangular' calves and thighs, art-connoisseurs might refuse to accept these as aesthetically valuable. They could hardly deny the analogy between limb-parts and wedges. But they would dismiss it as uninteresting, even ugly. And I explained: 'In their view, it is one thing to allow a human artist to challenge our perceptions, and upset our comfortable aesthetic conventions, but quite another to tolerate such impertinence from a computer program.'

What has impertinence to do with it? Well, to be impertinent is to say something (perhaps something both true and relevant) which one has no right to say. A person's right to be heard depends largely on their social status and topic-specific authority. These distinctions are necessary, because we cannot attend equally to everything that anyone says about anything.

At a certain level of generality, however, everyone has some authority. We all have aims, fears and beliefs, each of which – unless the contrary can be specifically shown – deserves to be respected. Everyone has a right to be heard, a right to try to persuade others and a right to further their interests. Each of us has these rights merely by virtue of being a member of the human community.

And that is the point. Computers are not automatic members of the human community, in the way that members of the biological species *homo sapiens* are. If they are not even members of our community, then they have none of our rights. So 'impertinence' makes sense.

However, automatic membership is only one way of entering a community: someone can be invited to join. It is up to the community to decide who – or what? – is acceptable.

Prima facie, the science-fictional computers we are discussing would have a strong claim to honorary membership in human conversational groups. For they could do many humanlike things – sometimes better than us. Many social functions – story-teller, jazz-musician, financial adviser, marriage-counsellor, psychotherapist – could be carried out by these non-biological 'intelligences'. So we might decide to do away with the scare-quotes entirely, when using psychological words to describe them.

But this decision, to acknowledge computers as *really* intelligent, would have far-ranging social implications. It would mean that, up to a point, we should consider their interests – much as we consider the interests of animals. For interests, *real* interests, they would be assumed to have.

Many current programs set up goals and sub-goals, and try to achieve them in various ways. (It is because we discovered that the hoverfly patently does not do this that we refuse to credit it with intelligence.) The highly advanced systems we are imagining would be able to do this too – and to ask for our cooperation. Suppose a computer-poet, tussling with a new composition, requested a good analogy for winter, or asked you to check whether thrushes' eggs are blue. If you had accepted it as a genuinely intelligent creature, you would be bound, within reason, to interrupt what you were doing in order to oblige it. (Even insects sometimes benefit from our fellow-feeling. You might squash an irritating hoverfly without much compunction, but would you deliberately pull a hoverfly to pieces, just for fun? And have you never got up out of your comfortable chair to put a ladybird into the garden?) I leave it to you to imagine scenarios in which more problematic conflicts of interest might occur.

Similarly, to regard computer-systems as *really* intelligent would mean that they could be deceived – and that, all things being equal, we should not deceive them. It would mean, too, that they could *really* know the things they were apparently saying, so we could *really* trust them.

These two examples have already come up in the English law courts. Ironically, the court's refusal in both cases to acknowledge real deception or knowledge seems unsatisfactory. The first example (which involved a mechanical device, not a computer) concerned a man who had lifted the 'arm' of a car-park machine without putting any money in. The magistrate acquitted him, on the grounds that to commit fraud one must deceive someone, and 'a machine cannot be deceived'. In the other example, the prisoner was accused of stealing banknotes. The prosecution submitted a list of banknote-numbers, some of which matched notes found in his possession. In law, documents accepted as evidence must be produced by someone 'having knowledge of' their contents. But the crucial list had been produced by the bank's computer. Because a computer (so the judge said) cannot have any knowledge of anything, the accused was acquitted.

My aim in giving those examples was not to start you on a life of crime, nor even to show that the law is an ass. The law will presumably be changed in some way, so as to prevent such absurdities. But just how should it be changed? Would you advise today's judges to accept that computers *can* have knowledge of the documents they produce? What about the judges living in the futuristic society we have been imagining? Would you be happy for one of that society's 'creative' computers to adjudicate such tricky legal decisions?

Whatever your answers, the crucial point is that the decision to remove all scare-quotes, when describing programs in psychological terms, carries significant moral overtones. So, like moral decisions in general, it cannot be forced upon us by the facts alone.

To answer 'No!' to the fourth Lovelace-question is to insist that, no matter how impressive future computers may be, we must retain all the moral and epistemological authority, and all the responsibility too. There can be no question of negotiating with computer-programs, no question of accepting – or even rejecting – their advice: the role of *adviser* is barred to them. Quasi-intelligent, quasi-creative programs would be widely used, much as pocket-calculators are today. But it would be up to us to take the entire responsibility for relying on their 'knowledge', and for trusting their 'advice'.

Whether people actually would answer 'No!', in the imaginary situation we are discussing, is not certain. Their answer might even hang on mere superficialities. For our moral attitudes and general sympathies are much influenced by biologically based factors, including what the other person – or quasi-person – looks like, sounds like and feels like.

Fur or slime, cuddliness or spikiness, naturally elicit very different responses. Walt Disney profited from the universal human tendency to caress and protect small animals with extra-large heads and extra-large eyes. Even robots can profit from this tendency. You may remember the two robots in the film *Star Wars*: the life-sized golden tin-man, CP3O, and the little, large-headed R2D2. At one point in the story, R2D2 toddles after CP3O as fast as he can on his two little legs, calling to him in his squeaky voice 'Wait for me!' When I saw the film, a chorus of indulgent 'Oohs' and 'Ahs' rose spontaneously from the cinema audience around me. So if our futuristic computers were encased in fur, given attractive voices and made to look like teddy-bears, we might be more morally accepting of them than we otherwise would be.[6] If they were made of organic materials (perhaps involving connectionist networks constructed out of real neurones), our moral responses might be even more tolerant.

You may have little patience with this science-fictional discussion. You may feel that we cannot know, now, what we would do in such a hypothetical situation. You may argue that we cannot even imagine it clearly, never mind decide what a morally appropriate response would be.

If so, I sympathize with you. The fourth Lovelace-question is, in large part, a disguised call for a complex moral–political decision concerning a barely conceivable situation. I have offered firm answers to the first three Lovelace-questions. As for the fourth, it can be left undecided.

Science is widely believed to destroy not only romance (converting the wonderful into the prosaic), but freedom as well. The deck-chairs of creativity lie alongside those of freedom: if the tide covers one, it covers all.

Blake's reference to freedom, in the passage cited above, is only one of many expressions of this view. Over a century before, on realizing that mankind is but a slender reed in the terrifying spaces of the newly conceived scientific universe, Blaise Pascal had consoled himself by noting that it is a *thinking* reed – and part of what he had in mind was our ability to make free choices. Our subjectivity, our ability to think of things and choose actions that have never been, and to construct the world of the mind over and above the external environment, was for him our saving glory. This, at least, lay beyond the cold impersonal touch of science.

But freedom, too, can be touched by science without being destroyed by it. A computational psychology can allow that much human action is self-generated and self-determined, involving deliberate choices grounded in personal loyalties and/or moral principles. Indeed, it shows us how free action is possible at all.

To a significant extent, arguments defending freedom against science parallel arguments pitting creativity against a scientific psychology. Worries about predictability and determinism, for example, crop up constantly in such discussions. The arguments detailed in Chapter 9 with respect to creativity apply, *pari passu*, to freedom too.

In either case, pure indeterminism gives mere chaos. To choose one's actions freely is not to hand responsibility over to the unpredictable. Luke Rhinehart's novel *The Dice Man* shows how unfree, and unhuman, a life would be in which most choices are determined by chance. And in either case, predictability may be a virtue. One would not go for advice on a moral problem to someone whose judgments are not usually reliable, nor admire someone whose good deeds were always grounded in passing whims.

Sensible, responsible thinking requires a highly structured conceptual space, and a discriminating exploration of the possibilities involved. Thoughts triggered by chance (R-random) events, in the environment or within the brain, may be fruitfully integrated into the psychological structures of the mind concerned. If so, all well and good. If not, they are either mere idle irrelevancies or (sometimes) tyrants forcing us to venture beyond the space of our self-determination.

An obsessional idea is a tyrant inside the mind. It monopolizes the person's attention, bypassing many of the self-reflective computations involved in genuinely free action. Acting responsibly involves careful deliberation, not enthusiastic spontaneity.

You may recall William Golding's remark that some incidents in his novels came to him rather than from him: 'I heard it. . . . [At such moments] the author becomes a spectator, appalled or delighted, but a spectator.' One might 'hear' a recommendation for action, too: many people have. But if matters of any moral importance are concerned, one should think hard before allowing it. People who claim to be obeying inner voices are surrendering their responsibility to someone supposedly more worthy (this is one way in which a tyrannical idea can bypass the normal deliberative channels). For Joan of Arc, the authority-figures were St Margaret and St Catherine; for the mass-murderer known as the Yorkshire Ripper, the voices seemed to come directly from God. (Hypnosis puts temporary tyrants into the mind; but hypnotists can rarely, if ever, force people to act in ways that conflict with their most fundamental evaluative principles.)

There are tyrants outside the mind, too. Someone who threatens you with imprisonment, or who holds a gun to your head, gives you such an overwhelming (even urgent) reason for doing what they command that your normal reasoning can get no purchase. You are not so constrained as the theatre-goers who, because of some computational short-cut in their minds, run without thinking when they hear someone behind them shout 'Fire!' You have more freedom than them, because you do have the computational capacity (especially in a non-urgent situation) to choose to disobey the tyrant. But the sanctions involved are so extreme, and your fear probably so great, that you are most unlikely to choose the action you know to be right. What in normal circumstances is unremarkable, in these situations would be heroic.

Free choice is structured choice, not mere mental coin-tossing. Even the *acte gratuit* of the existentialist (as depicted by Albert Camus in *L'Étranger*, for instance) is purposefully generated, to make a philosophical point. But the philosophy concerned is mistaken. Human freedom is not wholly unconditioned, any more than creative 'intuition' is.

Our ability to perform apparently random, motiveless acts is undeniable – but only if they are done with some purpose or other in mind do they count as *acts* in the first place. Someone who kicks the cat while having an epileptic fit is not only not acting freely: their kicks were not actions at all. Nor are the hoverfly's journeyings to meet its mates. Actions must be generated by certain sorts of computational structure (goal-hierarchies, for example) in the mind. They have a complex psychological grammar, whereas epileptic kicks do not. Camus' hero, who commits suicide as a defiant gesture in face of what he sees as the absurdity of the world, is indeed acting freely. That is, he is making a fundamentally self-determined choice, as opposed to following unexamined habits and unquestioned principles.

The person who acts out of 'bad faith' does not take the trouble to explore – still less, to transform – the conceptual space within which their habitual actions are situated. Sartre's waiter plays the waiter's role (follows the waiter-script), without ever considering any other possibilities. Someone who does not even ask 'What else could I do?' is rather like the nineteenth-century chemist who takes it for granted that all molecules must be strings. The question of their not being strings does not even arise, for how *could* they be anything else? (It does not follow that we should constantly raise questions about every aspect of our habitual behaviour; as noted in Chapter 5, to dispense with role-scripts entirely would be to drown in computational overload.)

Choices that arouse our world-weary cynicism are like hack novels, written in an undemanding literary style and displaying a shoddy sense of priorities. Choices that take us, with some subtlety, down previously unexplored pathways already marked on familiar moral maps are like the pleasing improvisations of a jazz-musician who can play in only one style. Choices that lead to fundamentally unexpected actions, arousing a shock of admiration or contempt, are comparable to Kekulé's creative transformation of an accepted chemical constraint. And someone who changes the basic topography of our moral landscape (for instance, by considering the negative and saying 'Love your *enemy*') is like someone who composes music in a radically new style.

All these are examples of free choice. But, as George Orwell might have put it, some are more free than others. We have seen throughout this book that it is usually unhelpful to ask, 'Is that idea creative: *yes* or *no*?', because creativity exists in many forms, and on many levels. Much of the interest, and the illumination, lies in the details. So, too, it is usually unhelpful to ask, 'Was that action free: *yes* or *no*?' Action in general, of which the generation (and evaluation) of new ideas is a special case, has a highly complex psychological structure. What may count as 'free' (or 'creative') in light of one of the many relevant structural aspects may count as relatively 'unfree' (or 'unimaginative') in light of others.

A person's psychological structure includes their myriad beliefs, goals, anxieties, preferences, loyalties and moral–political principles – in a word, their subjectivity. Each of us is different from every other human being, and these differences influence what we do, and how we think, in systematic ways. We all construct our lives from our own idiosyncratic viewpoints.

In Newton's vision, however, we are all alike. Indeed, the magnificence of his theory lay in its ability to see apples, tides and planets – and human bodies, too – as instances of one unifying principle. Even the clinical anatomist or biochemist, who makes distinctions between one person and another, is not professionally interested in *personal* distinctions. The natural sciences (including neurophysiology) are not concerned with personal, subjective phenomena. These cannot even be described in natural-scientific vocabulary – and what cannot be described is very likely to be ignored, or even denied. It is hardly surprising, then, that science *as such* is widely seen as inevitably dehumanizing.

It does not follow that *no* science can admit our individual differences, or explain them in terms of general principles of mental function. But the concepts used in such a science must be able to describe ideas, and subjective thought. They must be able to depict the mental processes that generate our idiosyncratic representations of the world, including our ideas of other people's minds and meanings.

This, as we have seen, is what a computational psychology can do. It cannot hope to rival the novelist's eye for human detail, or the poet's insight into human experience. But that is not its job. The task of a scientific psychology is to explain, in general terms, *how such things are possible*.

Over a century ago, people were asking how the diversity of biological species, and the layered order within the fossil record, are possible. Darwin criticized the cramped imagination of those who could wonder only at piecemeal special creations. Understanding the body's laws of harmony, he said, provided a more magnificent view.

Similarly, to attribute creativity to divine inspiration, or to some unanalysable power of intuition, is to suffer from a paucity of ideas. Even to describe it as the bisociation of matrices, or as the combination of old concepts in new ways, does not get us very far. But, surprising though it may seem, our imagination can be liberated by a computational psychology. The explanatory potential of this approach is much – very much – more than most people imagine.

The Loom of Locke and the Water-Wheels of Newton had no room for notions like creativity, freedom and subjectivity. As a result, the matters of the mind have been insidiously downgraded in scientific circles for several centuries. It is hardly surprising, then, if the myths sung by inspirationists and romantics have been music to our ears. While science kept silent about imagination, anti-scientific songs naturally held the stage.

Now, at last, computational psychology is helping us to understand such things in scientific terms. It does this without lessening our wonder, or our self-respect, in any way. On the contrary, it increases them, by showing how extraordinary is the ordinary person's mind. We are, after all, humans – not hoverflies.

References

Chapter 1: The Mystery of Creativity

1 A. Koestler, *The Act of Creation* (London, 1975), p. 211.
2 A. Lovelace, 'Notes on Manabrea's Sketch of the Analytical Engine Invented by Charles Babbage', in B. V. Bowden (ed.), *Faster Than Thought* (London, 1953), p. 398; see also A. Hyman, *Charles Babbage: Pioneer of the Computer* (Oxford, 1982).
3 L.A. Lerner, *A.R.T.H.U.R.: The Life and Opinions of a Digital Computer* (Hassocks, 1974).

Chapter 2: The Story So Far

1 Quoted in Koestler, *The Act of Creation*, p. 117.
2 Quoted in A. Findlay, *A Hundred Years of Chemistry* (London, 1965), p. 39.
3 Quoted in ibid., pp. 38–9.
4 W. Golding, *The Hot Gates* (London, 1965), p. 98.
5 W. Hildesheimer, *Mozart* (London, 1983), p. 15.
6 J. Livingston Lowes, *The Road to Xanadu: A Study in the Ways of the Imagination* (London, 1951), p. 358.
7 H. Poincaré, *The Foundations of Science: Science and Hypothesis, The Value of Science, Science and Method* (Washington, 1982), p. 389.
8 Ibid., pp. 390–1.
9 Ibid., p. 393.
10 Ibid., p. 386.
11 Koestler, *The Act of Creation*, p. 210.
12 Ibid., p. 121.
13 Ibid., p. 201.
14 D.N. Perkins, *The Mind's Best Work* (Cambridge, Mass., 1981).
15 H.E. Gruber, *Darwin on Man: A Psychological Study of Scientific Creativity* (London, 1974).
16 M. Polanyi, *Personal Knowledge: Towards a Post-Critical Philosophy (New York, 1964)*.
17 Quoted in Koestler, *The Act of Creation*, p. 117.
18 Quoted in ibid., p. 170.
19 Quoted in ibid., p. 117.
20 Quoted in Livingston Lowes, *The Road to Xanadu*, p. 498.
21 Koestler, *The Act of Creation*, p. 217.

22 Ibid., pp. 391–2.

Chapter 3: Thinking the Impossible

1 Quoted in Koestler, *The Act of Creation*, p. 120.
2 G. Taylor, *Reinventing Shakespeare* (London, 1990).

Chapter 4: Maps of the Mind

1 D.R. Hofstadter, *Godel, Escher, Bach: An Eternal Golden Braid* (New York, 1979).
2 Quoted in Findlay, *A Hundred Years of Chemistry*, p. 39.
3 G. Polya, *How to Solve It: A New Aspect of Mathematical Method* (Princeton, 1945).
4 S. Papert, *Mindstorms: Children, Computers, and Powerful Ideas* (Brighton, 1980); E. de Bono, *De Bono's Thinking Course* (London, 1982).
5 C. Rosen, *Schoenberg* (Glasgow, 1976).
6 T.S. Kuhn, *The Structure of Scientific Revolutions* (Chicago, 1962).
7 A. Karmiloff-Smith, 'Constraints on Representational Change: Evidence from Children's Drawing', *Cognition*, 34 (1990), pp. 57–83.
8 A. Karmiloff-Smith, 'From Meta-processes to Conscious Access: Evidence from Children's Metalinguistic and Repair Data', *Cognition*, 23 (1986), pp. 95–147.
9 Ibid., pp. 77–8.

Chapter 5: Concepts of Computation

1 H.C. Longuet-Higgins, *Mental Processes: Studies in Cognitive Science* (Cambridge, Mass., 1987), part II.
2 R.C. Shank and R.P. Abelson, *Scripts, Plans, Goals, and Understanding* (Hillsdale, NJ, 1977).
3 R.C. Schank and P. Childers, *The Creative Attitude: Learning to Ask and Answer the Right Questions* (New York, 1988).
4 H.L. Gelernter, 'Realization of a Geometry-Theorem Proving Machine', in E.A. Feigenbaum and J. Feldman (eds), *Computers and Thought* (New York, 1963), pp. 134–52.

Chapter 6: Creative Connections

1 F. Jacob, *The Statue Within: An Autobiography* (New York, 1988), p. 296.
2 Livingston Lowes, *The Road to Xanadu*.
3 D.E. Rumelhart and J.L. McClelland (eds), *Parallel Distributed Processing: Explorations in the Microstructure of Cognition* (Cambridge, Mass., 1986). The chapter on 'Distributed Representations' is reprinted in M.A. Boden (ed.), *The Philosophy of Artificial Intelligence* (Oxford, 1990), ch. 11.
4 Rumelhart and McClelland, *Parallel Distributed Processing*, vol. I, ch. 7.
5 Ibid., vol. 2, ch. 18.
6 S. Pinker and A. Prince, 'On Language and Connectionism: Analysis of a Parallel Distributed Processing Model of Language Acquisition', *Cognition*, 28 (1988), 73–193; A. Clark, *Microcognition: Philosophy, Cognitive Science, and Parallel Distributed Processing* (London, 1989), ch. 9.

Chapter 7: Unromantic Artists

1 M. Sharples, *Cognition, Computers, and Creative Writing* (Chichester, 1985).
2 Three exhibition-catalogues are: *Harold Cohen: Drawing* (San Francisco Museum of Modern Art, 1979); *Harold Cohen* (Tate Gallery, 1983); *Harold Cohen: Computer-as-Artist* (Buhl Science Center, Pittsburgh, 1984). See also H. Cohen, 'On the Modelling of Creative Behavior' (Santa Monica, 1981); H. Cohen, 'How to Make a Drawing' (1982).
3 In *Harold Cohen: Computer-as-Artist.*
4 P.N. Johnson-Laird, *The Computer and the Mind: An Introduction to Cognitive Science* (London, 1988), ch. 14; P.N. Johnson-Laird, 'Freedom and Constraint in Creativity', in R.J. Sternberg (ed.), *The Nature of Creativity: Contemporary Psychological Perspectives* (Cambridge, 1988), pp. 202–19; P.N. Johnson-Laird, 'Jazz Improvisation: A Theory at the Computational Level' (unpublished working-paper, 1989).
5 Johnson-Laird cites Parsons' *Directory of Tunes and Musical Themes* (1975) in 'Jazz Improvisation', p. 31.
6 M. Masterman, 'Computerized Haiku', in J. Reichardt (ed.), *Cybernetics, Art, and Ideas* (London, 1971), pp. 175–83; M. Masterman and R. McKinnon Wood, 'Computerized Japanese Haiku', in J. Reichardt (ed.), *Cybernetic Serendipity* (London, 1968), pp. 54–5.
7 Described in M.A. Boden, *Artificial Intelligence and Natural Man* (London, 1987), pp. 299–304, 312–14.
8 M.G. Dyer, *In-Depth Understanding: A Computer Model of Integrated Processing for Narrative Comprehension* (Cambridge, Mass., 1983).
9 R.C. Schank and C.K. Riesbeck (eds), *Inside Computer Understanding: Five Programs Plus Miniatures* (Hillsdale, NJ, 1981), pp. 197–258.
10 R.P. Abelson, 'The Structure of Belief Systems', in R.C. Schank and K.M. Colby (eds), *Computer Models of Thought and Language* (San Francisco, 1973), pp. 287–340.
11 K. Oatley and P.N. Johnson-Laird, 'Towards a Cognitive Theory of the Emotions', *Cognition and Emotion*, 1 (1987), pp. 29–50; P.N. Johnson-Laird and K. Oatley, 'The Language of Emotions: An Analysis of a Semantic Field', *Cognition and Emotion*, 3 (1989), pp. 81–123.
12 A. Davey, *Discourse Production: A Computer Model of Some Aspects of a Speaker* (Edinburgh, 1978).
13 K.J. Holyoak and P. Thagard, 'Analogical Mapping by Constraint Satisfaction', *Cognitive Science*, 13 (1989), pp. 295–356.
14 P. Thagard, K.J. Holyoak, G. Nelson and D. Gochfeld, 'Analog Retrieval by Constraint Satisfaction' (unpublished research-paper, 1988).
15 Koestler, *The Act of Creation*, p. 201.
16 D.R. Hofstadter, *Metamagical Themas: Questing for the Essence of Mind and Pattern* (London, 1985), chs 13 and 24.

Chapter 8: Computer-Scientists

1 R.S. Michalski and R.L. Chilausky, 'Learning by Being Told and Learning from Examples: An Experimental Comparison of the Two Methods of Knowledge Acquisition in the Context of Developing an Expert System for Soybean Disease

Diagnosis', *International Journal of Policy Analysis and Information Systems*, 4 (1980), pp. 125–61.

2 D. Michie and R. Johnston, *The Creative Computer: Machine Intelligence and Human Knowledge* (London, 1984), pp. 110–12.

3 Ibid., pp. 122–5.

4 R. Lindsay, B.G. Buchanan, E.A. Feigenbaum and J. Lederberg, *DENDRAL* (New York, 1980); B.G. Buchanan, D.H. Smith, W.C. White, R. Gritter, E.A. Feigenbaum, J. Lederberg and C. Djerassi, 'Applications of Artificial Intelligence for Chemical Inference: XXII Automatic Rule Formation in Mass Spectrometry by Means of the Meta-Dendral Program', *Journal of the American Chemistry Society*, 98 (1976), pp. 6168–78.

5 J. Glanvill, *The Vanity of Dogmatizing: The Three 'Versions'* (Brighton, 1970).

6 P. Langley, H.A. Simon, G.L. Bradshaw and J.M. Zytkow, *Scientific Discovery: Computational Explorations of the Creative Process* (Cambridge, Mass., 1987).

7 D.B. Lenat, 'The Ubiquity of Discovery', *Artificial Intelligence*, 9 (1977), pp. 257–86; D.B. Lenat, 'The Role of Heuristics in Learning by Discovery: Three Case Studies', in R.S. Michalski, J.G. Carbonell and T.M. Mitchell (eds), *Machine Learning: An Artificial Intelligence Approach* (Palo Alto, Calif., 1983); D.B. Lenat and J. Seely Brown, 'Why AM and EURISKO Appear to Work', *Artificial Intelligence*, 23 (1984), pp. 269–94; G.D. Ritchie and F.K. Hanna, 'AM: A Case Study in AI Methodology', *Artificial Intelligence*, 23 (1984), pp. 249–68.

8 Lenat, 'The Role of Heuristics in Learning by Discovery'.

9 Described in J.H. Holland, K.J. Holyoak, R.E. Nisbett and P.R. Thagard, *Induction: Processes of Inference, Learning, and Discovery* (Cambridge, Mass., 1986), pp. 124–6.

10 Ibid.

11 B.S. Johnson, *Aren't You Rather Young to be Writing Your Memoirs?* (London, 1973), pp. 24–31.

12 An article on dice-music in *Musical Times* (October 1968) is cited on p. 154 of Michie and Johnston, *The Creative Computer*.

13 Holland *et al.*, *Induction*, ch. 11.

14 P. Thagard, 'Explanatory Coherence', *Behavioral and Brain Sciences*, 12 (1989), pp. 435–502.

Chapter 9: Chance, Chaos, Randomness, Unpredictability

1 A patient suffering from Tourette's syndrome, described by the neurologist Oliver Sacks in an essay in the *New York Review of Books*.

2 J. Gleick, *Chaos: Making a New Science* (London, 1988); I. Stewart, *Does God Play Dice?: The Mathematics of Chaos* (Oxford, 1989).

3 C.A. Skarda and W.J. Freeman, 'How Brains Make Chaos in Order to Make Sense of the World', *Behavioral and Brain Sciences*, 10 (1987), pp. 161–96.

4 Quoted in Livingston Lowes, *The Road to Xanadu*, p. 148.

Chapter 10: Elite or Everyman?

1 Schank and Childers, *The Creative Attitude*.

2 Perkins, *The Mind's Best Work*, p. 33.

3 D.C. Marr, *Vision* (San Francisco, 1982).

4 Quoted in Koestler, *The Act of Creation*, p. 329.
5 R.W. Weisberg, 'Problem Solving and Creativity', in R.J. Sternberg, *The Nature of Creativity*, p. 171.
6 J.R. Hayes, *The Complete Problem Solver* (Philadelphia, 1981).
7 Koestler, *The Act of Creation*, p. 240.
8 Papert, *Mindstorms*.
9 Ibid.; S. Weir, *Cultivating Minds: A LOGO Casebook* (New York, 1987).
10 R.D. Pea and D.M. Kurland, 'On the Cognitive Effects of Learning Computer Programming', *New Ideas in Psychology*, 2 (1984), pp. 137–68.
11 Oatley and Johnson-Laird, 'Towards a Cognitive Theory of the Emotions'; Johnson-Laird and Oatley, 'The Language of Emotions: An Analysis of a Semantic Field'; A. Sloman, 'Motives, Mechanisms, and Emotions', in Boden, *The Philosophy of Artificial Intelligence*, ch. 10; M.A. Boden, *Purposive Explanation in Psychology* (Cambridge, Mass., 1972), chs 5–7.
12 H. Gardner, *Frames of Mind: The Theory of Multiple Intelligences* (London, 1983), ch. 6.
13 J.H. Kunkel, 'Vivaldi in Venice: An Historical Test of Psychological Propositions', *Psychological Record*, 35 (1985), pp. 445–57.

Chapter 11: Of Humans and Hoverflies

1 Koestler, *The Act of Creation*, p. 391.
2 J.R. Searle, 'Minds, Brains, and Programs', reprinted in Boden, *The Philosophy of Artificial Intelligence*, ch. 3. (A fuller version of my reply is 'Escaping from the Chinese Room', in Boden, *The Philosophy of Artificial Intelligence*, ch. 4.)
3 A. Sloman, 'What Sorts of Machines Can Understand the Symbols They Use?', *Proceedings of the Aristotelian Society*, Supplementary Volume, 60 (1986), pp. 61–80.
4 C.H. Thigpen and H.M. Cleckley, *The Three Faces of Eve* (London, 1957).
5 T. Nagel, 'What is it Like to be a Bat?', *Philosophical Review*, 83 (1974), pp. 435–57.
6 N. Frude, *The Intimate Machine: Close Encounters with the New Computers* (London, 1983).

Bibliography

n.a. *Harold Cohen: Drawing*. San Francisco: San Francisco Museum of Modern Art, 1979.

n.a. *Harold Cohen*. London: The Tate Gallery, 1983.

n.a. *Harold Cohen: Computer-as-Artist*. Pittsburgh: Buhl Science Center, 1984.

Abelson, R.P. 'The Structure of Belief Systems', in R.C. Schank and K.M. Colby (eds), *Computer Models of Thought and Language*. San Francisco: Freeman, 1973, pp. 287–340.

Boden, M.A. *Purposive Explanation in Psychology*. Cambridge, Mass.: Harvard University Press, 1972.

Boden, M.A. *Artificial Intelligence and Natural Man*. London: MIT Press; New York: Basic Books, 1987. 2nd edn, expanded.

Boden, M.A. ed. *The Philosophy of Artificial Intelligence*. Oxford: Oxford University Press, 1990.

Buchanan, B.G., D.H. Smith, W.C. White, R. Gritter, E.A. Feigenbaum, J. Lederberg and C. Djerassi. 'Applications of Artificial Intelligence for Chemical Inference: XXII Automatic Rule Formation in Mass Spectrometry by Means of the Meta-Dendral Program', *Journal of the American Chemistry Society*, 98 (1976), pp. 6168–78.

Clark, A. *Microcognition: Philosophy, Cognitive Science, and Parallel Distributed Processing*. London: MIT Press, 1989.

Cohen, H. *On the Modelling of Creative Behavior*. Santa Monica, Calif.: Rand Corporation, 1981. Rand Paper p. 6681.

Cohen, H. *How to Make a Drawing*. Talk given to the Science Colloquium, National Bureau of Standards, Washington DC. 17 December 1982.

Davey, A. *Discourse Production: A Computer Model of Some Aspects of a Speaker*. Edinburgh: Edinburgh University Press, 1978.

De Bono, E. *De Bono's Thinking Course* London: BBC, 1982.

Dyer, M.G. *In-Depth Understanding: A Computer Model of Integrated Processing for Narrative Comprehension*. Cambridge, Mass.: MIT Press, 1983.

Findlay, A. *A Hundred Years of Chemistry*. 3rd edn, ed. T.I. Williams. London: Duckworth, 1965.

Frude, N. *The Intimate Machine: Close Encounters with the New Computers*. London: Century, 1983.

Gardner, H. *Frames of Mind: The Theory of Multiple Intelligences*. London: Heinemann, 1983.

Gelernter, H.L. 'Realization of a Geometry-Theorem Proving Machine', in E.A. Feigenbaum and J. Feldman (eds), *Computers and Thought*. New York: McGraw-Hill, 1963, pp. 134–52.

Glanvill, J. *The Vanity of Dogmatizing: The*

Three 'Versions'. Brighton: Harvester, 1970.

Gleick, J. *Chaos: Making a New Science*. London: Heinemann, 1988.

Golding, W. *The Hot Gates*. London: Faber & Faber, 1965.

Gruber, H.E. *Darwin on Man: A Psychological Study of Scientific Creativity*. London: Wildwood House, 1974.

Hayes, J.R. *The Complete Problem Solver*. Philadelphia: Franklin Institute Press, 1981.

Hildesheimer, W. *Mozart*. London: Vintage, 1983.

Hofstadter, D.R. *Godel, Escher, Bach: An Eternal Golden Braid*. New York: Basic Books, 1979.

Hofstadter, D.R. *Metamagical Themas: Questing for the Essence of Mind and Pattern*. London: Viking, 1985.

Holland, J.H., K.J. Holyoak, R.E. Nisbett and P.R. Thagard. *Induction: Processes of Inference, Learning, and Discovery*. Cambridge, Mass.: MIT Press, 1986.

Holyoak, K.J. and P. Thagard. 'Analogical Mapping by Constraint Satisfaction', *Cognitive Science*, 13 (1989), pp. 295–356.

Hyman, A. *Charles Babbage: Pioneer of the Computer*. Oxford: Oxford University Press, 1982.

Jacob, F. *The Statue Within: An Autobiography*. New York: Basic Books, 1988.

Johnson, B.S. *Aren't You Rather Young to be Writing Your Memoirs?* London: Hutchinson, 1973.

Johnson-Laird, P.N. *The Computer and the Mind: An Introduction to Cognitive Science*. London: Fontana, 1988.

Johnson-Laird, P.N. 'Freedom and Constraint in Creativity', in R.J. Sternberg (ed.), *The Nature of Creativity: Contemporary Psychological Perspectives*. Cambridge: Cambridge University Press, 1988, pp. 202–19.

Johnson-Laird, P.N. 'Jazz Improvisation: A Theory at the Computational Level', unpublished working-paper, MRC Applied Psychology Unit, Cambridge, 1989.

Johnson-Laird, P.N. and K. Oatley. 'The Language of Emotions: An Analysis of a Semantic Field', *Cognition and Emotion*, 3 (1989), 81–123.

Karmiloff-Smith, A. 'From Meta-processes to Conscious Access: Evidence from Children's Metalinguistic and Repair Data', *Cognition*, 23 (1986), pp. 95–147.

Karmiloff-Smith, A. 'Constraints on Representational Change: Evidence from Children's Drawing', *Cognition*, 34 (1990), 57–83.

Koestler, A. *The Act of Creation*. London: Picador, 1975. (First published 1964.)

Kuhn, T.S. *The Structure of Scientific Revolutions*. Chicago: University of Chicago Press, 1962.

Kunkel, J.H. 'Vivaldi in Venice: An Historical Test of Psychological Propositions', *Psychological Record*, 35 (1985), pp. 445–57.

Langley, P., H.A. Simon, G.L. Bradshaw and J.M. Zytkow. *Scientific Discovery: Computational Explorations of the Creative Process*. Cambridge, Mass.: MIT Press, 1987.

Lenat, D.B. 'The Ubiquity of Discovery', *Artificial Intelligence*, 9 (1977), pp. 257–86.

Lenat, D.B. 'The Role of Heuristics in Learning by Discovery: Three Case Studies', in R.S. Michalski, J.G. Carbonell and T.M. Mitchell (eds), *Machine Learning: An Artificial Intelligence Approach*. Palo Alto, Calif.: Tioga, 1983.

Lenat, D.B. and J. Seely Brown. 'Why AM and EURISKO Appear to Work', *Artificial Intelligence*, 23 (1984), pp. 269–94.

Lerner, L.A. *A.R.T.H.U.R.: The Life and Opinions of a Digital Computer*. Hassocks, Sussex: Harvester Press, 1974.

Lindsay, R., B.G. Buchanan, E.A. Feigenbaum and J. Lederberg. *DENDRAL*. New York: McGraw-Hill, 1980.

Livingston Lowes, J. *The Road to Xanadu: A Study in the Ways of the Imagination.* London: Constable, 1951. (2nd edition.)

Longuet-Higgins, H.C. *Mental Processes: Studies in Cognitive Science.* Cambridge, Mass.: MIT Press, 1987.

Lovelace, A. 'Notes on Manabrea's Sketch of the Analytical Engine Invented by Charles Babbage', in B.V. Bowden (ed.), *Faster Than Thought.* London: Pitman, 1953, pp. 362–408.

Marr, D.E. *Vision.* San Francisco: Freeman, 1982.

Masterman, M. 'Computerized Haiku', in J. Reichardt (ed.), *Cybernetics, Art, and Ideas.* London: Studio Vista, 1971, pp. 175–83.

Masterman, M. and R. McKinnon Wood. 'Computerized Japanese Haiku', in J. Reichardt (ed.), *Cybernetic Serendipity.* London: Studio International, 1968, pp. 54–5.

Michalski, R.S. and R.L. Chilausky. 'Learning by Being Told and Learning from Examples: An Experimental Comparison of the Two Methods of Knowledge Acquisition in the Context of Developing an Expert System for Soybean Disease Diagnosis', *International Journal of Policy Analysis and Information Systems,* 4 (1980), pp. 125–61.

Michie, D. and R. Johnston. *The Creative Computer: Machine Intelligence and Human Knowledge.* London: Viking, 1984.

Nagel, T. 'What is it Like to be a Bat?' *Philosophical Review,* 83 (1974), pp. 435–57.

Oatley, K. and P.N. Johnson-Laird. 'A Cognitive Theory of Emotions', *Cognition and Emotion,* 1 (1987), pp. 29–50.

Papert, S. *Mindstorms: Children, Computers, and Powerful Ideas.* Brighton: Harvester Press, 1980.

Pea, R.D. and D.M. Kurland. 'On the Cognitive Effects of Learning Computer Programming', *New Ideas in Psychology,* 2 (1984), pp. 137–68.

Perkins, D.N. *The Mind's Best Work.* Cambridge, Mass.: Harvard University Press, 1981.

Pinker, S. and A. Prince, 'On Language and Connectionism: Analysis of a Parallel Distributed Processing Model of Language Acquisition', *Cognition,* 28 (1988), pp. 73–193.

Poincaré, H. *The Foundations of Science: Science and Hypothesis, The Value of Science, Science and Method.* Washington: University Press of America, 1982.

Polanyi, M. *Personal Knowledge: Towards a Post-Critical Philosophy.* New York: Harper, 1964.

Polya, George. *How to Solve It: A New Aspect of Mathematical Method.* Princeton, NJ: Princeton University Press, 1945.

Ritchie, G.D. and F.K. Hanna. 'AM: A Case Study in AI Methodology', *Artificial Intelligence,* 23 (1984), pp. 249–68.

Rosen, C. *Schoenberg.* Glasgow, Collins, 1976.

Rumelhart, D.E. and J.L. McClelland (eds). *Parallel Distributed Processing: Explorations in the Microstructure of Cognition.* 2 vols. Cambridge, Mass.: MIT Press, 1986.

Schank, R.C. and R.P. Abelson. *Scripts, Plans, Goals, and Understanding.* Hillsdale, NJ: Erlbaum, 1977.

Schank, R.C. and P. Childers, *The Creative Attitude: Learning to Ask and Answer the Right Questions.* New York: Macmillan, 1988.

Schank, R.C. and K.M. Colby (eds), *Computer Models of Thought and Language.* San Francisco, 1973.

Schank, R.C. and C.K. Reisbeck (eds). *Inside Computer Understanding: Five Programs Plus Miniatures.* Hillsdale, NJ: Erlbaum Press, 1981.

Searle, J.R. 'Minds, Brains, and Programs', *Behavioral and Brain Sciences,* 3 (1980), pp. 473–97. Reprinted in M.A. Boden (ed.), *The Philosophy of Artificial*

Intelligence. Oxford: Oxford University Press, 1990, ch. 3.

Sharples, M. *Cognition, Computers, and Creative Writing*. Chichester: Ellis Horwood, 1985.

Skarda, C.A. and W.J. Freeman. 'How Brains Make Chaos in Order to Make Sense of the World', *Behavioral and Brain Sciences*, 10 (1987), pp. 161–96.

Sloman, A. 'What Sorts of Machines Can Understand the Symbols They Use?' *Proceedings of the Aristotelian Society*, Supplementary Volume, 60 (1986), pp. 61–80.

Sloman, A. 'Motives, Mechanisms, and Emotions', *Cognition and Emotion*, 1 (1987), pp. 217–33. Reprinted in M.A. Boden (ed.), *The Philosophy of Artificial Intelligence*. Oxford: Oxford University Press, 1990, ch. 10.

Sternberg, R.J. (ed.). *The Nature of Creativity: Contemporary Psychological Perspectives*. Cambridge: Cambridge University Press, 1988.

Stewart, I. *Does God Play Dice?: The Mathematics of Chaos*. Oxford: Blackwell, 1989.

Taylor, G. *Reinventing Shakespeare*. London: Hogarth Press, 1990.

Thagard, P. 'Explanatory Coherence', *Behavioral and Brain Sciences*, 12 (1989), pp. 435–502.

Thagard, P., K.H. Holyoak, G. Nelson and D. Gochfeld. 'Analog Retrieval by Constraint Satisfaction'. Research-paper, Cognitive Science Laboratory, Princeton University. November, 1988.

Thigpen, C.H. and H.M. Cleckley. *The Three Faces of Eve*. London: Secker & Warburg, 1957.

Weir, S. *Cultivating Minds: A LOGO Casebook*. New York: Harper & Row, 1987.

Weisberg, R.W. 'Problem Solving and Creativity', in R.J. Sternberg (ed.), *The Nature of Creativity: Contemporary Psychological Perspectives*. Cambridge: Cambridge University Press, 1988, pp. 148–76.

Index